Autonomous Technology

Autonomous Technology

Technics-out-of-Control as a Theme in Political Thought

Langdon Winner

The MIT Press
Cambridge, Massachusetts, and London, England

This book was set by To the Lighthouse Press, printed in the United States of America.

Second printing, 1980

First MIT Press paperback edition, 1978

Library of Congress Cataloging in Publication Data

Winner, Langdon
Autonomous Technology.

Includes Index
1. Technology—Social aspects. 2. Technology—Philosophy. 3. Technocracy. I. Title.
T14.5.W56 301.24'3 76-40100
ISBN 0-262-23078-X (hardcover)
ISBN 0-262-73049-9 (paperback)

To my parents and Mrs. A

Contents

Preface

The questions that inform this book spring from my coming to terms with a set of problems in political and academic life of the past decade. My concerns took shape at an institution, the University of California at Berkeley, which had, for reasons its leaders thought eminently progressive, defined itself as a vast research, training, and social service mechanism. As it pursued that end, its philosophy began to exemplify a mode of human relationships that I here describe as technological politics. That model eventually came to grief in a situation that cast major social issues of the 1960s and early 1970s—civil rights, Vietnam, racism, and militarism—at the door of the "multiversity." In the lively and sometimes violent controversies of that period, Berkeley became a place in which politics in the plaza and learning in the classroom were never far removed. This was especially true for those of us fortunate to be studying political theory and intellectual history during those years. For whatever one's position on the war or latest university controversy happened to be, the lectures, sections, and readings on Thucydides, Aristotle, Hobbes, Locke, or Marx always seemed directly suited to dilemmas immediately at hand.

I wish to express special thanks to my teachers at Berkeley—Sheldon Wolin, John Schaar, Hanna Pitkin, Norman Jacobson, and Thomas Morrison—whose ability to mix scholarly concerns with broader human commitments is an inspiration to all who study with them. My involvement with issues of political thought, which they encouraged, was later given focus by the questions about technology and politics raised by Todd La Porte, a man whose interest in my work made it possible for me to continue when I might otherwise have stopped. Sheldon Wolin read the manuscript in its earliest stages and made many helpful suggestions about the direction it might take. Later drafts were considerably improved by the comments of Darrell Hawthorne, Jim Miller, Ira Kurzban, Todd La Porte, Robert Biller, and my colleagues at M.I.T., Miles Morgan, Charles Weiner, Nathan Sivin, and Joseph Weizenbaum. To Hanna Pitkin, Kai Lee, and Herman van Gunsteren I owe a special debt for their line-by-line commentaries on my writing. This would be a much better, albeit much longer book if I had followed all of their suggestions. As it is, I thank them for steering me through a number of difficulties and toward a project that now extends far beyond the confines of this one statement. Don Van Vliet, who does not read manuscripts, influenced much of what appears here through his sense of the comedy of nature and artifice. The friendship and encouragement of Greil Marcus and Bruce Miroff have often sustained me in this and other work.

What the reader finds here is a detailed treatment of a theme relevant to political theory, which, it seems to me, has never received sufficient attention. Ideological presuppositions in radical, conservative, and liberal thought have tended to prevent discussion of a problem in technics and politics that has forced itself onto the agenda of our times. The problem was first and most powerfully raised for me by a speaker on the steps of Sproul Hall in December 1964. "There is a time," his voice rang out, "when the operation of the machine becomes so odious, makes you so sick at heart that you can't take part; you can't even passively take part, and you've got to put your bodies upon the gears and upon the wheels, upon the levers, upon all the apparatus and you've got to make it stop. And you've got to indicate to the people who run it, to the people who own it, that unless you're free, the machine will be prevented from working at all."[1]

Those words marked a turning point for the student generation of that decade and gave me the first glimmerings of my subject here. Concealed in them was a contradiction, which events of subsequent years have made clear, that there are institutions one must oppose and struggle to modify even though one also has considerable affection for them.

Autonomous Technology

Introduction

In one disguise or another, technology has been a central theme in political thought for the past two hundred years. Although the definition of the issue of concern has again and again shifted, it has been clear during this time that there is something in the nature of modern technology thinkers can ill afford to ignore. A partial catalog of the topics that have been associated with various aspects of modern technics would include the following: the industrial revolution and the rise of industrial society, the ascendancy of the middle class, the possibility of utopia, the misery of the working class and the necessity of revolution, the rise of new elites, the social and psychological turmoil involved in rapid change, alienation, nationalism, imperialism, leisure, and the possibility of ecological disaster.

Despite its widely acknowledged importance, however, technology itself has seldom been a primary subject matter for political or social inquiries. While technological developments are commonly cited as among the most important causes of the shape of modern society, the tendency has been to see the matter solely in terms of economics and economic history, perspectives that due to their special mode of abstraction and selectivity give us a very limited vision of the role technics have played in modern history. Writers who have suggested the elevation of technology-related questions to a more central position—William F. Ogburn, Lewis Mumford, Leslie White, and others—have for the most part been politely ignored. The prevalent opinion has remained that the true problems of modernity could best be understood in ways that excluded all direct reference to the technical sphere. Technology could be left to the technicians.

In recent years, however, the prevailing winds of neglect have begun to shift. Technology and its various manifestations have become virtual obsessions in discussions about politics and society on a wide variety of fronts. Social scientists, politicians, bureaucrats, corporate managers, radical students, as well as natural scientists and engineers, are now united in the conclusion that something we call "technology" lies at the core of what is most troublesome in the condition of our world.

There is, of course, little agreement as to the nature of the problem or about the approach that an intelligent person should take in the quest for understanding. In the eyes of scientists and technicians, the issue takes the form of a moral dilemma that hovers menacingly over their work. Since World War II they have become increasingly sensitive to the fact that scientific technologies have profound and often unfortunate consequences in the world at large. With the neutrality of their professions and products now in question, they have begun intensive inquiries into the political and ethical context in which their activities exist.[1]

From the point of view of social scientists and managers, the crucial issues are those of the increasing complexity and rate of change in modern society. Developments in the technical sphere continually outpace the capacity of individuals and social systems to adapt. As the rate of technological innovation quickens, it becomes increasingly important and increasingly difficult to predict the range of effects that a given innovation will have. When compounded by the increasing complexity of sociotechnical systems, these changes make it more and more difficult to carry out some of the most basic activities of contemporary social life: planning, design, and functional coordination. For this reason complexity and change are increasingly studied as "independent variables" said to have objectively knowable correlations to certain kinds of social and political phenomena.[2]

In other modes of interpretation, however, the concerns of the natural and social scientists are held to be trivial, self-serving, and beside the point. Radical critics of "the technological society" in both Europe and America have insisted that what deserves our attention is not the rate of technological innovation and its effects but rather the very existence of advanced technology in the life of man. Technology is, according to this view, a source of domination that effectively rules all forms of modern thought and activity. Whether by an inherent property or by an incidental set of circumstances, technology looms as an oppressive force that poses a direct threat to human freedom. In the words of Allen Ginsberg, "Ourselves caught in the giant machine are conditioned to its terms, only holy vision or technological catastrophe or revolution break 'the mind-forg'd manacles.' "[3] A slogan of the Black Panther party, "The spirit of the people is greater than the man's technology," expresses the conviction that someday the system of domination will be overcome, a testable hypothesis somewhat different from those social scientists currently ponder.[4]

Technology is a word whose time has come. Its rise as a conscious problem in a wide variety of social and political theories requires some explanation. We are now faced with an odd situation in which one observer after another "discovers" technology and announces it to the world as something new. The fact is, of course, that there is nothing novel about technics, technological change, or advanced technological societies. One can argue that medieval Europe was a highly sophisticated technological society of a certain sort, involved in a fairly rapid, continuing process of sociotechnical change. One does not have to wait for the industrial revolution or the so-called postindustrial period of the twentieth century to see political societies remolded in response to technical innovation. We are justified in asking, then, why this topic should suddenly arise as a matter of intense concern.

Paul Goodman once suggested that the widespread uneasiness about science and technology amounts to a religious upheaval similar to that of the Protestant Reformation. "Science," he explained, "has long been the chief orthodoxy of modern times. And precisely science which should have been the wind of truth to clear the air, has polluted the air, helped to brainwash, and provided the weapons of war."[5] Current protests surrounding the military-industrial complex are in effect a call for a return to "the high tradition of science and technology" much like Luther's call for a new affirmation of the true Christian faith. A reformed scientific technology would reemphasize the ideals which, according to Goodman, once guided its progress: prudence, decentralization, ecology, and incorruptibility. Lewis Mumford holds much the same view. There is, he believes, a humane tradition of science and technology based on "an earth-centered, organic, and human model" to which Western civilization must return if it is to avoid the disastrous course of the "megamachine."[6] "For its effective salvation," Mumford warns, "mankind will need to undergo something like a spontaneous religious conversion: one that will replace the mechanical world picture, and give to the human personality, as the highest manifestation of life, the precedence it now gives to machines and computers."[7]

While such analogies of religious crisis help to illuminate the outrage present in much of the contemporary criticism of technology, they fail to capture an important characteristic of the discussion—its pervasive sense of puzzlement and disorientation. The writers who have isolated technology as an issue have repeatedly stressed that what is involved is not merely a problem of values or faith but, more importantly, a problem in our understanding of things. There is, they assert, something wrong in the way we view technology and man's relationship to it. In its present array of vast and complex forms, technology continually surprises us and baffles our attempts at comprehension. From all sides one hears the call for new evidence and new interpretations to remedy our disoriented state.

In this regard I would suggest that we supplement Goodman's New Reformation with what may be a more appropriate historical analogy, the scientific revolution. In the sixteenth and seventeenth centuries, after hundreds of years of relatively stable scientific belief, the realm of nature was suddenly cast open to question. The discoveries of Copernicus, Galileo, Kepler, Vesalius, and others placed all of God's creations in a new and surprising light—a light that inspired generations of inquiry into natural phenomena and resulted in a totally new conception of the physical universe and man's place within it. As Thomasso Campanella poignantly expressed it, "If Galilei's conclusions are right . . . we shall have to philosophize in a new way."[8]

In much the same manner the realm of technology has become an open question for the present age. After centuries in which technical artifice was of little interest outside the confines of its own development and practice, the nature of man's own creations has now emerged as a source of genuine perplexity. The technological world that the scientific revolution helped bring into being has itself become a focus of new inquiry. The crucial insight which occasions this new awakening is nothing so profound as the disclosure of Copernicus's *De Revolutionibus* that the earth revolves around the sun rather than vice versa. It is instead roughly equivalent to the realization that the sun rises in the morning and sets in the evening; for the astonishing fact that one thinker after another has stumbled upon is merely this: technology in its various manifestations is a significant part of the human world. Its structures, processes, and alterations enter into and become part of the structures, processes, and alterations of human consciousness, society, and politics. The remarkable impact of Marshall McLuhan and Jacques Ellul rests on their ability to sensitize modern audiences to something they had overlooked: we are surrounded on all sides (possibly even the inner side) by a myriad of techniques and technologies. Apparently these influences had become so much a part of everyday life that they had become virtually invisible. The changes and disruptions that an evolving technology repeatedly caused in modern life were accepted as given or inevitable simply because no one bothered to ask whether there were other possiblities. It is for this reason that the discussion about the place of technology in human existence requires much more than facile talk about how well or how poorly technology accords with "human values."[9] One can paraphrase Campanella in saying that if the observations of Ellul, McLuhan, Marcuse, Mumford, Sypher, Galbraith, and others are correct, we shall have to do at least some of the work of social science and political theory in a new way.

The analogy here, like most other analogies, is valid only if taken in moderation. In mentioning the scientific revolution in the same breath as our present questions about technology, I am not asking the reader to trace out all conceivable similarities between a science of nature and a science of artifice. In particular, I am not suggesting that the issues here are solely "empirical" ones that can be handled through improved social scientific methodology. The tendency in research of that kind is to define all problems as those of "change" and to gather data relevant to selected correlations. While such work is sometimes interesting, I have never been convinced that the crucial questions at hand are best studied in terms of "change." If there were never another technological breakthrough, innovation, or advance, and never another social, ecological, or political consequence, we would still face a host of problems about the meaning of technology in the life of man. Much of social scientific research in this area amounts to a triumph of instrumentation —virtuosity in measuring and comparing quantifiable variables—rather than an earnest effort to advance our understanding.

But it is not clear that we would know what to do with the new models and data even if we had them. Where does one encounter a rich and lively discussion about the practical, moral, and political context in which these findings make sense? Almost nowhere. The hope is that a new study of technological affairs would bring together the relevant spheres of knowledge, judgment, and action in a way that might point to more intelligent choices. Technology, after all, is inherently pragmatic. It deals with establishing what one wants and how one wants to pursue it. But in almost every book or article on the subject the discussion stalls on the same sterile conclusion: "We have demonstrated the relationship between Technology X and social changes A, B, and C. Obviously, Technology X has implications for astounding good or evil. It is now up to mankind to decide which the case will be."

Poor mankind. Although freshly equipped with the best findings of social science, it is still left holding the bag. At this point the fact-value distinction, considered as a moral imperative, has its most lethal effects. The social scientist, presumably the person who knows most about the issues at hand, ceases to inquire into the practical implications of his own work. To go further, he believes, is to tread on the soil of "values," an area that he holds to be little more than a tortuous field of personal preferences, prejudice, and half-brained moralism. The idea that there could be a reasonable basis upon which one could arrive at general conclusions about wise or unwise choices for political society is totally foreign to him. After he explains the relationships found in the data, his contribution ends.

The truth of the matter is that our deficiency does not lie in the want of well-verified "facts." What we lack is our bearings. The contemporary experience of things technological has repeatedly confounded our vision, our expectations, and our capacity to make intelligent judgments. Categories, arguments, conclusions, and choices that would have been entirely obvious in earlier times are obvious no longer. Patterns of perceptive thinking that were entirely reliable in the past now lead us systematically astray. Many of our standard conceptions of technology reveal a disorientation that borders on dissociation from reality. And as long as we lack the ability to make our situation intelligible, all of the "data" in the world will make no difference.

A good illustration of this state of disorientation can be seen in the peculiar way in which the word *technology* appears in academic and everyday speech. In past decades the term had a very specific, limited, and unproblematic meaning. Persons who employed the term spoke of a "practical art," "the study of the practical arts," or "the practical arts collectively." In the literature of the eighteenth and nineteenth centuries, such meanings were clear and were not the occasion for deliberation or analysis. *Technology*, in fact, was not an important term in descriptions of that part of the world we would now call technological. Most people spoke directly of machines, tools, factories, industry, crafts, and engineering and did not worry about "technology" as a distinctive phenomenon.

In the twentieth century, however, the linguistic convention has gradually changed. *Technology* has expanded rapidly in both its denotative and connotative meanings. It is now widely used in ordinary and academic speech to talk about an unbelievably diverse collection of phenomena—tools, instruments, machines, organizations, methods, techniques, systems, and the totality of all these and similar things in our experience. The shift in meaning from something relatively precise, limited, and unimportant to something vague, expansive, and highly significant can be traced through the definitions in Webster's unabridged dictionary. In *Webster's Second International* (1909) the word is said to mean "industrial science, the science or systematic knowledge of the industrial arts, especially of the more important manufactures." In *Webster's Third New International* (1961), however, the definition blossomed into the following: "the totality of means employed by a people to provide itself with the objects of material culture." Today, even this definition seems too narrow, for if we notice how the word is actually employed, it certainly covers much more than just the material objects of culture. Some of the most intriguing new technologies have to do with the alteration of psychological or spiritual states.

Many persons find it uncomfortable to leave the meaning of *technology* in this form. Social scientists usually insist that a precise, manageable operational definition be hammered out. From their point of view, if this is not done we will surely find ourselves in the position of Jacques Ellul who defines his central concept, *la technique*, as "*the totality of methods rationally arrived at and having absolute efficiency* (for a given stage of development) in *every* field of human activity."[10] Such a definition, Ellul's critics complain, is overly broad and does not approach the meaning of our word *technology*. I disagree. While Ellul's addition of "absolute efficiency" may cause us difficulties, his notion of *technique* as the totality of rational methods closely corresponds to the term *technology* as now used in everyday English. Ellul's *la technique* and our *technology* both point to a vast, diverse, ubiquitous totality that stands at the center of modern culture. Both include a substantial portion of what we make and what we do.

There is, of course, nothing unusual in the discovery that an important term is ambiguous or imprecise or that it covers a wide diversity of situations. Wittgenstein's discussion of "language games" and "family resemblances" in *Philosophical Investigations* illustrates how frequently this occurs in ordinary language. For many of our most important concepts, it is futile to look for a common element in the phenomena to which the concept refers. "*Look and see* whether there is anything common to all.—For if you look at them you will not see something that is common to all, but similarities, relationships, and a whole series of them at that."[11]

What is interesting in this case, however, is that a concept that was once very specific in the way it was used has now become amorphous in the extreme. There is a tendency among those who write or talk about technology in our time to conclude that technology is everything and everything is technology. In a dialectic of concepts that Hegel would have appreciated, the word has come to mean everything and anything; it therefore threatens to mean nothing.

For those who would listen to language rather than perform elaborate operations on it, this annoying symptom will not be taken as an occasion to impose an arbitrary definition. It should be seen as an interesting sign. What does this chaotic use of the term *technology* indicate to us?

An answer to this question is that while the sphere of technics one wishes to talk about has grown rapidly, the linguistic resources of public discourse have changed little at all. Specialists in the various subdivisions of technology have developed concepts to make their own sphere of activity intelligible to them; but for the most part these concepts remain foreign and even mysterious to the nonspecialist or the specialist of another field. The same concepts useful in building and maintaining a given technology are not those useful in understanding its broader implications for the human community. In this sense the confusion surrounding the concept "technology" is an indication of a kind of lag in public language, that is, a failure of both ordinary speech and social scientific discourse to keep pace with the reality that needs to be discussed. "Technology," therefore, is applied haphazardly to a staggering collection of phenomena, many of which are recent additions to our world. One feels that there must be a better way of expressing oneself about these developments, but at present our concepts fail us.

One consequence of this state of affairs is that discussions of the political implications of advanced technology have a tendency to slide into a polarity of good versus evil. Because there is no middle ground for talking about such things, statements often end up being expressions of total affirmation or total denial. One either hates technology or loves it. In my own attempts to speak with scientists, engineers, and managers over the years, I have again and again run into responses that refuse to tolerate any ambiguity on this cherished, threadbare dichotomy. I have tried to point out that America has for too long substituted technical solutions for problems that were either political or moral in nature. I have suggested that there might be some desirable alternatives to the ways in which we now employ various kinds of technology—for example, other ways of structuring the use of television than our present nationwide, corporate-owned networks. As innocuous as these views are, they are often taken as a threat. Any criticism of sociotechnical practice could only be vile opposition. "You're just using technology as a whipping boy," the response comes back. "You just want to stop progress and send us back to the Middle Ages with peasants dancing on the green."

A typical response of engineers, for example, is to announce that they are merely problem solvers. "Tell us the problem," they demand. "We will find a solution. That's our job. But you may not presume to question the nature of our solution. You are not a member of a technical profession and, therefore, know nothing of relevance. If you insist on raising questions about the appropriateness of the means we devise, we can only conclude that you are antitechnology."

It soon becomes clear that in this enlightened age there is almost no middle ground of rational discourse, no available common language with which persons of differing backgrounds can discuss matters of technology in thoughtful, critical terms. Conversations gravitate toward warring polarities and choosing sides. One source of fascination in my inquiries has been that existing discussions are often thoroughly nervous, even hysterical. When intelligent persons can become so upset over such ostensibly mundane matters, there is something peculiar going on.

It is not possible to clear up the inadequacies in our speech habits with a single stroke. But I shall offer some basic distinctions that I will be using in my writing here.

First I want to note the class of objects we normally refer to as technological—tools, instruments, machines, appliances, weapons, gadgets—which are used in accomplishing a wide variety of tasks. In speaking of objects of this sort I shall employ the term *apparatus*. For many persons, "technology" actually means apparatus, that is, the physical devices of technical performance.

I also want to mark the whole body of technical activities—skills, methods, procedures, routines—that people engage in to accomplish tasks and include such activities under the rubric *technique*. The root of this word is the Greek *technē* ("art," "craft," or "skill"), which linguists have further traced to the Indo-European root *teks-* ("to weave or fabricate"). From the earliest times, technique has been distinguished from other modes of human action by its purposive, rational step-by-step way of doing things.

In addition "technology" frequently refers to some (but not all) varieties of social organization—factories, workshops, bureaucracies, armies, research and development teams, and the like. For my uses here, the term *organization* will signify all varieties of technical (rational-productive) social arrangements. Another closely related term—*network*—will mark those large-scale systems that combine people and apparatus linked across great distances.

I am not a lexicographer and do not wish to legislate usage. These distinctions represent a modest attempt to bring a measure of order to a conversation that has lacked order so far, an attempt the rest of the book will continue. With this preliminary groundwork taken care of, let us turn to the central theme guiding our inquiries.

Chapter 1
Autonomy and Mastery

So the whole question comes down to this: can the human mind master what the human mind has made?
—Paul Valéry

One symptom of a profound stress that affects modern thought is the prevalence of the idea of autonomous technology—the belief that somehow technology has gotten out of control and follows its own course, independent of human direction. That this notion is (at least on the surface) patently bizarre has not prevented it from becoming a central obsession in nineteenth- and twentieth-century literature. For some time now, the writings of many of our most notable poets, novelists, scientists, and philosophers have been haunted by the fear that somehow technology has "run amok," is "no longer guided by human purposes," is "self-directing," or has "escaped all reasonable limits." Often occurring as a frank confession in the writings of an otherwise levelheaded, well-respected individual, the vision of technology out-of-control has taken a wide variety of forms. In his book *Physics and Philosophy*, Werner Heisenberg looks back over his work in quantum physics and concludes that he has unwittingly contributed to the rise of an uncontained historical force. "The enormous success of this combination of natural and technical science," he writes, "led to a strong preponderance of those nations or states or communities in which this kind of activity flourished, and as a natural consequence this activity had to be taken up even by those nations which by tradition would not have been inclined toward natural and technical sciences. The modern means of communication and of traffic finally completed this process of expansion of technical civilization. Undoubtedly the process has fundamentally changed the conditions of life on earth; and whether one approves of it or not, whether one calls it progress or danger, one must realize that it has gone far beyond any control through human forces. One may rather consider it as a biological process on the largest scale whereby the structures active in the human organism encroach on larger parts of matter and transform it into a state suited for the increasing human population."[1]

In John Kenneth Galbraith's *The New Industrial State* the notion appears as a stern warning to the American public. "I am led to the conclusion, which I trust others will find persuasive," Galbraith avers, "that we are becoming the servants in thought, as in action, of the machine we have created to serve us."[2] In *So Human an Animal,* René Dubos, the noted biologist, offers a view that combines conviction and total incredulity: "Technology cannot theoretically escape from human control, but in practice it is proceeding on an essentially independent course."[3] "Planning for better defined and worthwhile human goals has become urgent if we are to avoid the technological take-over and make technology once more the servant of man instead of his master."[4] Martin Heidegger, in *Discourse on Thinking,* asserts that the process has moved far beyond any possible repeal: "No one can foresee the radical changes to come. But technological advance will move faster and faster and can never be stopped. In all areas of his existence, man will be encircled ever more tightly by the forces of technology. These forces, which everywhere and every minute claim, enchain, drag along, press and impose upon man under the form of some technical contrivance or other—these forces . . . have moved long since beyond his will and have outgrown his capacity for decision."[5]

Until recently, concerns of this sort were the exclusive property of a small segment of the academic community. But during the last several years the idea of autonomous technology has gained considerable public attention. Observations to the effect that technology is "out of control" are commonly found in books atop the best-seller lists in the United States. The same opinion has become a cornerstone of the popular movement that marches forward under the ambiguous banner of "ecology." The notion is abroad in the land that "the forces of technology and commodity, allowed to have their own way without guidance or control of intervening values have created a culture which is profoundly hostile to life."[6] Technological innovation, it is said, has become so rapid and pervasive that it threatens to destroy all vestiges of permanence, continuity, and security in modern society. "Technology must be tamed, if the accelerative thrust is to be brought under control."[7]

The newly emergent vogue of the idea makes it clear that, if nothing else, "autonomous technology" has enough intuitive plausibility to stand as a convenient receptacle for a host of contemporary anxieties. In his memoirs Albert Speer, Hitler's minister of armaments and war production, finds it possible to hold up runaway technology as a partial excuse for the barbarism of the Nazi regime. "The criminal events of those years," Speer observes, "were not only an outgrowth of Hitler's personality. The extent of the crimes was also due to the fact that Hitler was the first to be able to employ the implements of technology to multiply crime."[8] Quoting from his testimony at the Nuremberg trial, Speer continues, "The more technological the world becomes, the greater is the danger. . . . As the former minister in charge of a highly developed armaments economy it is my last duty to state: A new great war will end with the destruction of human culture and civilization. There is nothing to stop unleashed technology and science from completing its work of destroying man which it has so terribly begun in this war."[9]

Given the historical context in which it stands, Speer's attempted apology does not succeed.[10] The interesting fact is, nonetheless, that so peculiar a defense could be offered at all and that it has been accepted so willingly by Speer's European and American audience. Something we all understand about the twentieth century makes the plea of Albert Speer and Adolf Eichmann, "I am a humble victim of autonomous technology," an intelligible, albeit still unjustifiable excuse.

In the present discussion the term *autonomous technology* is understood to be a general label for all conceptions and observations to the effect that technology is somehow out of control by human agency. My use of this notion stems most directly from Jacques Ellul's autonomous *technique*. According to Ellul, "Technique has become autonomous; it has fashioned an omnivorous world which obeys its own laws and which has renounced all tradition."[11] The theories I will examine here all maintain, in one way or another, that far from being controlled by the desired and rational ends of human beings, technology in a real sense now governs its own course, speed, and destination.

The concept of autonomy is particularly expressive in this context. Ellul is by no means the only person to have found a significant use for it in describing the technological society. Bruno Bettelheim has written of the threat to individual autonomy in a mass age,[12] while Galbraith warns of the apparent autonomy of the "technostructure" in the new industrial state. "Autonomy" is at heart a political or moral conception that brings together the ideas of freedom and control. To be autonomous is to be self-governing, independent, not ruled by an external law or force. In the metaphysics of Immanuel Kant, autonomy refers to the fundamental condition of free will—the capacity of the will to follow moral laws which it gives to itself. Kant opposes this idea to "heteronomy," the rule of the will by external laws, namely the deterministic laws of nature.[13] In this light the very mention of autonomous technology raises an unsettling irony, for the expected relationship of subject and object is exactly reversed. We are now reading all of the propositions backward. To say that technology is autonomous is to say that it is nonheteronomous, not governed by an external law. And what is the external law appropriate to technology? Human will, it would seem. But if technology can be shown to be nonheteronomous, what does this say about human will? Ellul is explicit on this point: "There can be no human autonomy in the face of technical autonomy."[14] In his eyes there is a one-for-one exchange.

The rise of notions of autonomous technology in Western literature has, in fact, come side by side with frequent and enthusiastic attacks on the idea of human autonomy. Especially in the more extreme reaches of behaviorist psychology and cybernetics, the idea of a self-governing human will is now thought to be a tired anachronism. In B. F. Skinner's *Beyond Freedom and Dignity* we learn that "a scientific analysis of behavior dispossesses autonomous man and turns the control he has been said to exert over to the environment."[15] Skinner's work tries to disabuse us of the illusion of personal autonomy and its associated concerns, freedom and responsibility. The alternative to autonomous man in Skinner's writing is very clear. It is technology—in particular, the technology of behavior control. "In trying to solve the terrifying problems that face us in the world today," he observes, "we naturally turn to the things we do best. We play from strength, and our strength is science and technology."[16] "What is needed is a technology of behavior, but we have been slow to develop the science from which such a technology might be drawn."[17]

My aim in this work is to identify a variety of notions of autonomous technology, to examine their basic rationale, and to inquire into the problems they suggest. In this regard, autonomous technology is a methodological touchstone, a clue leading to a number of larger issues. In some views the perception of technology-out-of-control is associated with a process of change in which the human world is progressively transformed and incorporated by an expanding scientific technology. In others the perception focuses upon the behavior of large-scale technical systems that appear to operate and grow through a process of self-generation beyond human intervention. In others still, the matter is primarily that of individuals dwarfed by the complex apparatus surrounding them, which they must employ if they are to survive. I shall be asking: What is involved in these complaints? What, if anything, is their truth? Along the way we will tackle concepts that are among the most perplexing in all of modern thought—technological dynamism, determinism, historical drift, technical imperatives, technocracy, and the like.

There are some obvious objections to the approach I have adopted here. My commitment to a perspective which takes technology as its primary focus and technological autonomy as its fundamental puzzle will no doubt be cause for some misgivings. In *Utopia and Its Enemies,* for example, George Kateb scoffs at the idea "that somehow machines will develop a volition of their own, independent of their makers and come eventually to change roles with man and make men their servants."[18] Such views, he contends, reflect only fear and hatred of the machine, "amongst the stalest and most pervasive emotions of modern life."[19] Those who see the world in this way deny the utopian possibilities that modern technology offers mankind.

For different reasons, Seymour Melman criticizes the notion of autonomous military technology in his *Pentagon Capitalism.* In his eyes the idea can be only a political cul de sac. "I am uneasy," he observes, "about theories viewing man as the captive of his weapons. This is a self-defeating mode of understanding, rather different from identifying the top decision-makers and their mode of control. Men may be captives, but only of other men. The concept of man in the grip of a Frankenstein weapons system has a severely limiting effect on our ability to do anything about it, if that is desired."[20]

Criticisms of this sort make a great deal of sense. I would be the first to admit that the approach I have chosen is one-sided and that it excludes much that is important in political and social life. I would also allow that there are some very real dangers in the view that technology and science are autonomous. Such notions have in the past sometimes accompanied philosophies which were virulently antimodern and even fascist.[21]

My justification in taking the present approach is that there are some significant questions here that the more obvious ways of talking about politics and technology have been prone to ignore. One way of raising these issues from concealment is to begin by looking at them in their most vivid outlines. From symptomology we can then move towards a reasonable diagnosis. Much of the writing on politics and technology today begins with the rational, concrete signs and moves from there gradually toward the incredible specter of autonomous technology. My choice is to work backward, or at least differently, in the hope of arriving at a more intelligent destination.

Mastery and Its Loss

Our starting point is the fact that the idea of autonomous technology has recurrently gained adherents in many widely separated branches of learning. From the beginning of the nineteenth century to the present day, the notion has perplexed the likes of Thomas Carlyle, Charles Dickens, Ralph Waldo Emerson, Nathaniel Hawthorne, Henry Thoreau, Mark Twain, Henry Adams, John Ruskin, William Morris, E. E. Cummings, George Orwell, Marcel Duchamp, and Kurt Vonnegut, to name just a few. In one definition or another autonomous technology is now a significant transdisciplinary hypothesis in the natural and social sciences, the arts, journalism, and even the technical specialties themselves.

But how can this be? How can we account for the fact that so many intelligent persons have embraced an idea so strange and unlikely?

The common element in notions of this kind is not so much a feeling of ambivalence about modern technics or even a loss of faith in the link between technological development and human progress. It is, rather, a sense that many of our most fundamental expectations about the technical sphere no longer hold. Common sense and the traditional view of technics do not always provide a reliable guide to our everyday experience with technological phenomena. In this sense, reports of autonomous technology can be interpreted as signs of a disorder of the mind at the collapse of an ordinary point of view. Let us examine the traditional perspective and seek out its present vulnerability.

The conclusion that something is "out of control" is interesting to us only insofar as we expect that it ought to be in control in the first place. Not all cultures, for example, share our insistence that the ability to control things is a necessary prerequisite of human survival. There are peoples who have lived and prospered under the belief that an inherent harmony or beneficence in nature would provide for their needs. Western culture, however, has long believed that its continued existence and advancement depend upon the ability to manipulate the circumstances of the material world. In a spirit that many have called Faustian, we believe that control is possible and that we must strive for it. As a both necessary and noble aspect of Western self-identity, we strive to isolate the variable conditions of the environment and manipulate them for our own advantage.

The concern of science and technology with the possibilities of control have often found expression in terms which closely parallel the language of politics. This is perhaps not surprising if one recalls that both politics and technics have as their central focus the sources and exercise of power. Our thinking about technology, however, seems inextricably bound to a single conception of the manner in which power is used—the style of absolute mastery, the despotic, one-way control of the master over the slave. Other notions central to the historical discussion of political power—membership, participation, and authority founded on consent—seem to have no relevance in this sphere. In our traditional ways of thinking, the concept of mastery and the master-slave metaphor are the dominant ways of describing man's relationship to nature, as well as to the implements of technology.

Much of the existing literature, for example, holds that technology and the human slave are exact equivalents, even to the point that they are functionally interchangeable. Historians of science speculate that one reason why Greek technology never developed to its full potential was the presence of the institution of slavery in Greek city-states. According to E. J. Dijksterhuis, "The fact that living machines could be used at will, no doubt diminished the need for inanimate implements, the more so because no humanitarian consideration formed a motive for calling in the aid of such implements. If machines, however, were deemed superfluous because they were slaves . . . , a vicious circle must have been created: for want of machines the slave could not be dispensed with."[22] The virtual equivalence of slave and machine plays a prominent role in Aristotle's discussion of slavery in the *Politics*. Aristotle explains that the instruments available to a man who owns property are of two sorts—animate and inanimate. The animate instruments, human slaves, are "prior to other instruments" in the sense that they must be present before the inanimate instruments can be used. "There is only one condition," Aristotle comments, "in which we can imagine managers not needing subordinates, and masters not needing slaves. This condition would be that each [inanimate] instrument could do its own work, at the word of command or by intelligent anticipation, like the statues of Daedalus or the tripod made by Hephaestus, of which Homer relates that 'of their own motion they entered the conclave of Gods on Olympus,' as if a shuttle should weave of itself, and a plectrum should do its own harp-playing."[23]

Contemporary discussions of automation echo Aristotle's conclusion. It is now possible for inanimate instruments to perform their own work "at the word of a command or by intelligent anticipation," that is, by a computer program. This development has led to conjecture that the perfection of industrial technology will eventually liberate mankind from toil. The metaphor employed by those who express this hope is the same as that introduced 2,500 years ago: something must be enslaved in order that something else may win emancipation.

The theme of mastery in the literature of technology is even more evident with regard to Western man's relationship to nature. Here there are seldom any reservations about man's rightful role in conquering, vanquishing, and subjugating everything natural. This is his power and his glory. What would in other situations seem rather tawdry and despicable intentions are here the most honorable of virtues. Nature is the universal prey, to manipulate as humans see fit.

In no place is this theme more clearly stated than in the writings of the most famous early advocate of a world-transforming scientific revlution, Francis Bacon. Book 1 of *Novum Organum* finds the philosopher laying plans for a new science that will advance man's understanding of nature. Criticizing the illusions, dogmas, and methods of inquiry of past philosophies, Bacon outlines a science that will combine rigorous observation, experiment, inductive logic, and organized scientific research. But although Bacon emphasizes the need for a new method in his treatment, he also insists that the new science will require new ends. "It is not possible," he observes, "to run a course aright when the goal itself has not been rightly placed. Now the true and lawful goal of the sciences is none other than this: that human life be endowed with new discoveries and powers."[24] Bacon returns to this topic later, announcing that it is time to "say a few words touching the excellency of the end in view."[25] His statement is an interesting one, for it establishes a direct comparison between the powers and goals of his new science and the powers and goals of politics. He seeks to demonstrate that scientific action is similar to political action, but that it is in all ways superior.

Bacon begins by noting that former ages had always awarded "divine honors" to those who made important discoveries or who were "authors of inventions." Persons of this stripe were esteemed far above "those who did good service in the state (such as founders of cities and empires, legislators, saviours of their country from endured evils, quellers of tyrannies, and the like)." To political actors, former ages "decreed no higher honour than heroic," and rightly so. "For the benefits of discoveries may extend to the whole race of man, civil benefits only to particular places; the latter not beyond a few ages, the former through all time."[26] Bacon notes that science and the practical arts had recently given the world three inventions—the printing press, gunpowder, and the magnet—that made all political accomplishments pale by comparison. "For these three have changed the whole face and state of things throughout the world . . . insofar that no empire, no sect, no star seems to have exerted greater power and influence in human affairs than these mechanical discoveries."[27]

The point of the matter is not merely that the action of science and technology brings a greater good over a greater area for a longer period of time; it is also superior to politics because it provides a more noble outlet for man's aggressive impulses. Here Bacon distinguishes between three kinds of ambition in mankind, two of which are political. "The first," he says, "is of those who desire to extend their own power in their native country; which kind is vulgar and degenerate. The second is of those who labour to extend the power of their country and its dominion among men. This certainly has more dignity, though not less covetousness."[28] The third kind of ambition, however, avoids these moral pitfalls. "If a man endeavors to establish and extend the power and dominion of the human race itself over the universe, his ambition (if ambition it can be called) is without doubt both a more wholesome thing and a more noble than the other two." Bacon leaves no doubt as to how dominion of this sort will be established: "Now the empire of man over things depends wholly on the arts and science. For we cannot command nature except by obeying her."[29]

What was previously a political impulse—the desire for power, conquest, and empire—shall now become the guiding impulse of science and technology. Men shall "obey" nature for as long as it takes to learn her secrets. They will then command her as tyrants once commanded their political subjects. Bacon clearly means to say that this change will benefit the human race not only because science will improve material well-being but also because those who crave power will turn to more "wholesome" pursuits. Apparently an ambitious man must subjugate something. And nature, unlike human beings, will not mind subjugation.

The success of Bacon's program as a way of knowing, as a vision of the world, and as a way of operating on material reality is perhaps the most important of all accomplishments in modern history. The terms of this accomplishment, to a large extent, define the conditions under which we live. This is all the more extraordinary given the fact that in earlier ambitious attempts, science and scientific technology had often been spectacular failures. In Chaucer's "Canon's Yeoman's Tale," for example, we read of a man (possibly Chaucer himself) who frittered away his fortune on a scientific art, alchemy, which never brought him any benefit at all:

When we had fixed a place to exercise
Our esoteric craft, we all looked wise;
Our terms were highly technical and quaint.
I blew the fire up til fit to faint. . . .

And how, d'you think? It happens, like as not,
There's an explosion and good-bye the pot!
These metals are so violent when they split
Our very walls can scarce stand up to it. . . .
Although the devil didn't show his face
I'm pretty sure he was about the place.
In Hell itself where he is lord and master

There couldn't be more rancour in disaster
Than when our pots exploded as I told you;
All think they've been let down and start to scold you. . . .

My face is wan and wears a leaden look;
If you try science you'll be brought to book.
My eyes are bleared with work on preparations,
That's all the good you get from transmutations.
That slippery science stripped me down so bare
That I'm worth nothing, here or anywhere.[30]

Incompetent, fumbling, and unproductive, the science of Chaucer's era was a waste of time and money. Up to and including the seventeenth century, science ran neck and neck with the quackery of the occult. Even the greatest figures of the scientific revolution still dabbled in the magical arts. Kepler was a confirmed astrologer; Newton tried his hand at alchemy. The triumph of the Baconian model of science, particularly in its technological applications, did not take place overnight. Its promise long preceded its fulfillment; its advertisements and advance men came long before its ability to deliver. Science in Newton's time was able to account for such things as the motion of the planets, but it had little direct effect on the practical life of society.[31] The enthusiasm of Diderot, d'Alembert, and the authors of the *Encyclopedia* for science and the mechanical arts was more a projection of things to come than a realistic assessment of the existing state of affairs. The consensus of historians of science and technology is that the wedding of science and technics came only in the late nineteenth century, with most of the progeny of this match arriving in the twentieth.[32]

Science, then, succeeded first as a way of knowing and as a vision of the world. Only later, as scientific technology, did it triumph as a means of control and manipulation. In the end, however, its ultimate success must be accounted to its fulfillment of Baconian ambitions—the delivery of power. Other modes of knowing have been able to give an intelligible, systematic, aesthetically pleasing picture of reality. If science had only been able to accomplish this and nothing more, it is likely that it would have been supplanted by yet another philosophy of inquiry. But in the West at least, the test is not so much *what do you know?* or *how elegant is your interpretation of worldly phenomena?* but rather, *what can you actually do?* This is the conclusive factor, the reason that, for instance, social science has never fully established its credentials in the halls of science. Science succeeds over rival ways of knowing—poetry, religion, art, philosophy, the occult—not by its ability to illuminate, not even by its ability to organize knowledge, but by its ability to produce solid results. Space scientists demonstrate that they *know* since they are able to fly a man to the moon and back. Physicists must *know* since they are able to produce atomic weapons. In the last analysis, the popular proof of science is technology. This is why we consider Bacon prophetic, Paracelsus quaint.

One importance of the idea of autonomous technology is that it sets out to debunk the dream of mastery by showing that it has gone awry in practice. In modern speculation about technics, one thinker after another has found it necessary to question the fundamental conceptions and beliefs that anchor the way Western men think about their relationship to technological possibilities. In specific, one finds a set of notions, once thought altogether reliable, which have become targets of widespread doubt. They are:

—that men know best what they themselves have made;

—that the things men make are under their firm control;

—that technology is essentially neutral, a means to an end; the benefit or harm it brings depends on how men use it.

In the conventional perspective works of technology are more than certain; they are doubly certain. Since human beings are both the designers and makers of their creations, they have precise knowledge of their construction. They know exactly how things are put together and how they can be taken apart. In addition, works of technology are certain in the sense that their construction depends on the possession of valid knowledge, either from mundane experience or from an appropriate science. In his *Metaphysics* Aristotle explains that *technē* ("art" or "craft") is superior to mere experience because it combines these two ways of knowing. "Art is born when out of the many bits of information derived from experience there emerges a grasp of these similarities in view of which they are a unified whole." "We believe that knowing and understanding characterize art rather than experience. . . . Men of experience discern the fact 'that,' but not the reason 'why'; whereas experts know the reason why and explanation."[33] Aristotle conjectures that the first man to invent an art or craft was looked upon with wonder by his fellows, not only because there was something useful in his discoveries but also because the others thought him wise and superior. Master workers are presumably wiser not merely because they are practical but because they have reasons and can explain what they are doing.

That which men have made, they also control. This is common sense. Control, after all, is part of the very design of technical creations. Apparatus and techniques are devised with definite purposes in mind. Through conscious manipulation of such means, men are able to achieve ends established in advance. While it may require a period of time to find instruments that will be effective, once discovered they are no longer a source of difficulty. Technical means are, by their very nature, mere tools subject to the will of whomever employs them. The fortuitous combination of certainty and control in technical activity has held a great fascination for politics. Many historical forms of statecraft and almost all conceptions of utopia rest on an implicitly technological model. When conditions of the political world seem uncertain, unmanageable, or otherwise undesirable, technique and artifice offer a tantalizing solution. Through conscious resort to artificial means it is conceivably possible, in the ironic words of Oscar Wilde, to "shield ourselves from the sordid perils of actual existence." If one could remake the world, if one could fashion the conditions of reality to suit a preconceived design, both certainty and control would be assured. This is the solution Thomas Hobbes offered: "For by art is created that great LEVIATHAN called a COMMONWEALTH, or STATE," a political order that would be more perfect since it was built from the ground up on entirely rational principles. In one manifestation or another, this is also the prescription presented us by Plato, Saint-Simon, Owen, Madison, and contemporary systems analysis.[34]

Technology is essentially neutral. In the conventional way of thinking, the moral context appropriate to technical matters is entirely clear. Technology is nothing more than a tool. What men do with tools, of course, is to "use" them. The tool itself is completely neutral—a means to the desired end. Whether the end accomplished is wise or unwise, beautiful or hideous, beneficial or harmful, must be determined independently of the instrument employed. This judgment also holds true for the wonderful developments in modern technology. The new devices, regardless of their size and complexity, are still tools that may be used either well or poorly. In the words of H. L. Nieburg, "Science and technology are essentially amoral and their uses ambivalent. Their miracle has increased equally the scale of both good and evil."[35] The neutrality of technology and the tool-use ethic are truisms striving to become bromides. This has not been an obstacle to the dozens of authors in the last decade or so who have unearthed these ideas and presented them as startling insights into the modern condition. In an attempt to correct the position that technology is intrinsically benevolent, while avoiding the notion that it is intrinsically evil, these authors have rediscovered the obvious.[36]

The three propositions I have outlined here are conspicuously reasonable. They are basic to any understanding of human mastery through technology. Yet the fact remains that much of modern discussion suggests that these ideas are no longer completely valid given the conditions of advanced technics.

How thoroughly do people know their own technology? This is an ambiguous question and could be answered in any number of ways. But if it means, How much does an individual understand about the total range of technologies that affect his or her life? the answer is clear. Very little. Technical knowledge in modern society is so highly specialized and diffused that most people can grasp only a minute segment. The rest of the technical activity and apparatus that surround each individual remain largely uncomprehended. Knowledge of how things are put together and how they work exceed the grasp of everyone other than the expert directly concerned with the particulars. The specialist, of course, is largely oblivious to the nature of the processes and configurations outside his field. Ideologies of the professions actually dignify this state of affairs by making a virtue of *not* knowing anything more than the particular segment. As Paul Goodman has observed, an indication of the noncomprehensibility of our technologies can be seen in the fact that most persons are totally helpless in matters of repair. When a complex mechanism breaks down, one must call in someone who understands its mysteries and can put it back in order. One meaning of the idea of mastery is that one is able to have a complete vision of something from beginning to end and complete facility in its use. In this sense, mastery in the technological society is increasingly rare. People work within and are served by technical organizations that by their very nature forbid a perspicuous overview. In this sense, complaints about autonomous technology are frequently of the sort: "I do not understand what is happening around me."

To what extent do men control technology? If control is understood to mean either the exercise of a dominating influence or holding in restraint, then much of modern literature would find the matter of control at best paradoxical. One now finds persistent depositions given about the following kinds of phenomena: large-scale systems that appear to expand by some inherent momentum or growth—weapons systems, freeways, skyscrapers, power, and communications networks—which make the notions of controlled application and reasonable use seem absurd; a continuing and ever-accelerating process of technical innovation in all spheres of life, which brings with it numerous "unintended" and uncontrolled consequences in nature and society; technical systems entirely removed from the possibility of influence through outside direction, which respond only to the requirements of their own internal operations. In other words, the same technologies that have extended man's control over the world are themselves difficult to control. Recognition of this fact is not limited to the critics of the technological society. It extends to the most highly developed sectors of big technology where the science of cybernetics and an obsession with "command and control" have been responses to internal difficulties of this sort. Another layer of sophisticated technology was required to enable the manipulation and coordination of networks already in use. As fragility and unpredictability in large-scale systems become facts of modern life, the ultimate success of these measures is still in doubt. Reports of autonomous technology, therefore, are sometimes of the sort: "The mechanism does not perform as expected; the slave will not obey."

Is technology a neutral tool to human ends? No longer can an affirmative answer be given without severe qualifications. The most spectacular of our implements often frustrate our ends and intentions for them. Skepticism greets the promise that "our transportation crisis will be solved by a bigger plane or a wider road, mental illness with a pill, poverty with a law, slums with a bulldozer, urban conflict with a gas."[37] Even the most distinguished of technocrats, Robert S. McNamara for example, experience the vexation of discovering that their mammoth technical systems cannot accomplish the basic tasks for which they were designed—winning a war, rebuliding the cities, and so forth. Although virtually limitless in their power, our technologies are tools without handles. Often they seem to resist guidance by preconceived goals or standards. Far from being merely neutral, our technologies provide a positive content to the area of life in which they are applied, enhancing certain ends, denying or even destroying others. Who would assert, for example, that the technologies of the industrial age were neutral with regard to the highest ideals of medieval society? Most important, our technical means sometimes take on the aspect of self-perpetuation or self-generation. Human beings still have a nominal presence in the network, but they have lost their roles as active, directing agents. They tend to obey uncritically the norms and requirements of the systems which they allegedly govern. Here a revaluation of values takes place in a manner that Nietzsche would have found abhorrent—through technical necessity. In the technicized operational definition of human ends, one often finds that a crucial element in the original notion has been lost and that something peculiar has been added in its place. The determinant of this strange addition (which soon becomes part of the life process) is usually a requirement of the technique employed. Often, then, our theme is announced when observers perceive that "the tools are much more than tools. Technological neutrality is a myth."

In summary, the loss of mastery manifests itself in a decline of our ability to know, to judge, or to control our technical means. It is in this general waning of intellectual, moral, and political command that ideas of autonomous technology find their basis. True, the acknowledgment of these symptoms is neither uniform nor universal. Persons who feel sufficiently empowered by the technical systems of our time will find such observations of little help. My hope is that there will be at least a few to whom the theme makes sense and who will want to consider it further. These, I suspect, will be persons who have already in some way experienced frustration in their attempts to make modern technics intelligible or accessible in their own lives. Perhaps even the most confident and best served will find reason to ponder the situation anew. I have never met any bureaucrat or expert who did not complain of having no real power or who gave any evidence of knowing much outside his or her own technical function. It is for those who can understand the modern significance of Heraclitus's maxim, "They are estranged from that with which they have most constant intercourse," that my investigations are intended.

Autonomy and Animism

One entrance to the problem, a kind of preview of the territory ahead, is to notice how it has been represented in works of art. During the last century and a half, the idea of autonomous technology has found expression in countless novels, poems, plays, and motion pictures. Standard to such treatments has been a mode of symbolism that portrays technological artifice as something literally alive. Through some strange process a man-made creature, machine, or advanced system takes on lifelike properties—consciousness, will, and spontaneous motion—which place it in rebellion against the human community.

An amusing recent example of this genre is a film, *Colossus: The Forbin Project,* which shows a brilliant scientist (obviously modeled on the figure of Wernher von Braun) and a liberal, technocratic U.S. President (who bears a striking resemblance to John F. Kennedy) plotting to jam and disarm their own giant computer, which has taken control of the economy and entire defense system of the United States and the Soviet Union. At the conclusion of the movie the computer intercedes, telling the hapless humans it has been amused by their fumbling efforts but that the game has gone on long enough. Colossus has the conspirators, including his creator, Dr. Forbin, locked up.

The often vulgar Hollywood use of technological animism should not obscure the fact that images of this kind have been useful symbols for artists and writers concerned with the implications of modern technical artifice. The root idea, of course, is nothing new. Its origins are as old as the tale of Prometheus fashioning the human race from clay, the story of man's creation in *Genesis,* or any number of similar accounts in world religions that depict the human species as an autonomous artifact of the gods. In this regard the notion of a living technology merely recapitulates the myths of our own beginnings—the rebellion and fall of man—and the ensuing harvest of troubles. In such writings as Mary Shelley's *Frankenstein* (which I shall discuss in some detail later) and Samuel Butler's *Erewhon,* the myth was powerfully reborn in Western thought. Nineteenth-century writers were fascinated by the possibility that scientists and inventors would actually succeed in creating artificial life. That prospect was important in its own right, but it was also symbolic of the growth of industrial society as well as more basic dilemmas involved in any creative act.[38] Artists have always understood that their works in a true sense "have a life of their own" both during and after their creation. When the advance of science and industrial technology became well known, artists were among the first to speculate on the possibilities for a truly perfect, independent, and lifelike work of human fabrication.

In Nathaniel Hawthorne's "The Artist of the Beautiful," for example, an inventor, Owen Warland, "considered it possible, in a certain sense, to spiritualize machinery, and to combine with the new species of life and motion thus produced a beauty that should attain to the ideal which Nature has proposed to herself in all her creatures, but has never taken pains to realize."[39] His friends, of course, scoff at his dream, but one day Warland presents them with his masterwork, a tiny mechanical butterfly of such supreme perfection as to be "alive." The creature flutters gracefully about the room as the friends gasp in amazement and the inventor discourses on the spiritual essence of beauty. But when the butterfly approaches the hand of his creator, it is denied. " 'Not so! not so!' murmured Owen Warland, as if his handiwork could have understood him. 'Thou hast gone forth out of thy master's heart. There is no return for thee.' "[40]

The possibility of perfection and, therefore, of superiority of technical creations was also the subject of an unusual piece of writing by Edgar Allan Poe. In 1835 Poe went to see a famous automaton, the "Chess Player," designed by the European inventor Johann Nepomuk Maelzel, the same Maelzel who had designed a mechanical brass orchestra, "The Panharmonicon," for Ludwig van Beethoven (who wrote "The Battle of Victoria" for the elaborate music box) and whose automated chess player had astounded Napoleon Bonaparte. The "Chess Player" was a mechanical man sitting behind a desk who played a very respectable, albeit jerky, game of chess. But Poe was not convinced. He studied the operation of the machine and decided, correctly it turns out, that Maelzel's device was a clever fake. In a long review of the show, Poe offers comprehensive proof that there is a man hidden in the mechanism. The clincher in his argument comes when he points out that, lo and behold, the "Chess Player" has been known to lose! "The Automaton does not invariably win the game," he explains. "Were the machine a pure machine, this would not be the case—it would always win."[41]

Poe's logic is direct but amusing. Since we design machines for perfection and superiority over human capacity, those that are not superior cannot be true machines. The practical implications of such technical possibilities became the focus of the animistic literature of the twentieth century. Automation is now much more than a speculative fantasy. Its capacity for the liberation from toil must be balanced against the prospect that man will find himself functionless in his own world. The prospect is much more than mere technological unemployment, unless one understands that notion in its most comprehensive sense. In the world of mechanical or electrochemical "dystopia" people would be left with absolutely nothing to do or be. So perfect is their work that they have fabricated themselves out of any meaningful existence. Those who can find a function at all will have to take on the character of their own robots, for anything that does not conform to the design of technological utopia cannot operate, much less have any utility.

Representative of contemporary fictional literature on this theme is Kurt Vonnegut's *Player Piano*. One of the book's subplots concerns the sad fate of Rudy Hertz, a former mechanic and hapless victim of automation. The day finally came when technicians had placed his whole job into the memory of the EPICAC computer. "Rudy hadn't understood quite what the recording instruments were all about, but what he had understood, he'd liked: that he, out of thousands of machinists, had been chosen to have his motions immortalized on tape."[42] Now, sitting in a tavern with his old friend, Dr. Paul Proteus, the protagonist in the novel, Rudy decides to put a coin in the player piano slot:

"I played this song in your honor, Doctor," shouted Rudy above the racket. "Wait till it's over." Rudy acted as though the antique instrument were the newest of all wonders, and he excitedly pointed out identifiable musical patterns in the bobbing keys—trills, spectacular runs up the keyboard, and the slow, methodical rise and fall of keys in the bass. "See—see them two go up and down, Doctor: Just the way the feller hit 'em. Look at 'em go!"

The music stopped abruptly, with the air of having delivered exactly five cents worth of joy. Rudy still shouted. "Makes you feel kind of creepy, don't it, Doctor, watching them keys go up and down? You can almost see a ghost sitting there playing his heart out."

Paul twisted free and hurried out to his car.[43]

Vonnegut's anecdote provides a glimpse of the crucial statement and ultimate conclusion of the writings on technological animism. If one asks, Where did this strange life in the apparatus come from? What is its real origin? the answer is clear: it is human life transferred into artifice. Men export their own vital powers—the ability to move, to experience, to work, and to think—into the devices of their making. They then experience this life as something removed and alien, something that comes back at them from another direction. In this way the experience of men's lives becomes entirely vicarious. And often it is full of surprises. In Hawthorne's story, a young woman demands of Owen Warland, "Tell me if it be alive, or whether you created it."[44] The inventor replies that both conclusions about the butterfly are equally true. "Alive? Yes, Annie; it may well be said to possess life, for it has absorbed my own being into itself."[45]

Much of the writing in this tradition suggests that there is a law of the preservation of life at work, much like the law of the preservation of energy. The transference is absolute; insofar as men pour their own life into the apparatus, their own vitality is that much diminished. The transference of human energy and character leaves men empty, although they may never acknowledge the void. This is the problem that the symbolism of such stories presents to us, one that is readily applicable to the dilemmas of autonomous technology. Man now lives *in* and *through* technical creations. The peculiar properties we may notice in these creations are not the result of some spontateous generation (the mistake of vulgar science fiction films). What we see is human life separated from the directing, controlling positive agency of human minds and souls.

This state of affairs would be of small interest if it had to do only with individual scientists and inventors and their works. In the most skillful statements of the theme, the specific piece of apparatus represents a much more general condition: the plight of the human race in technological society. This is clearly the case in E. M. Forster's classic story, "The Machine Stops," which describes a civilization living far beneath the surface of the earth, completely dependent upon a large, beneficent Machine. The Machine is both powerful and perfect, based on the principle that things must be brought to people, rather than people brought to things. "Those funny old days, when men went for change of air instead of changing the air in their rooms!"[46] The human beings, who are so well taken care of that they have shrunk in both mental and physical capacity, obey the Machine's every requirement. Only one person, a young man named Kuno, is perceptive enough to discern what has happened. He crawls to the surface of the earth through one of the Machine's vomitories and discovers that the land above is beautiful and that one does not have to rely on the Machine to breathe. When he returns to the caverns below he tells his friends: "Cannot you see . . . that it is we that are dying, and that down here the only thing that really lives is the Machine? We created the Machine, to do our will, but we cannot make it do our will now. It has robbed us of the sense of space and the sense of touch; it has blurred every human relation and narrowed down love to a carnal act, it has paralyzed our bodies and our wills, and now it compels us to worship it. The Machine develops—but not on our lines. The Machine proceeds—but not to our goal. We exist only as the blood corpuscles that course through its arteries, and if it could work without us, it would let us die."[47]

Writing in the late 1920s, Forster avoids the tendency of his time to see the future of technological society in terms of mechanization, the ordering of all social functions in monotonous, clocklike patterns. Instead, he is concerned with the rise of a pathological dependence that drains from men all activity and spirit. The life of the whole society is transferred into the mechanism. While the Machine returns what it has received in the form of "service," it retains the living essence that gave this "service" its original meaning. Here is a civilization in which sophisticated devices will think for people, work for people, improve their own workings for people, and in general provide for all of people's needs. The price paid for this wonder is that human beings can no longer have any direct relationship with each other or with the world; everything must go through the technical intermediary, and here something is always lost. "She fancied that he looked sad. She could not be sure, for the Machine did not transmit *nuances* of expression. It gave only a general idea of people—an idea that was good enough for all practical purposes."[48] Virtually unnoticed to itself and in a host of subtle ways, the human race and all of its cultural forms are dying. But those who built the machine and now live with its manifest conveniences have lost the ability to conceive of anything different. "Above her, beneath her, and around her, the Machine hummed eternally; she did not notice the noise, for she had been born with it in her ears."[49]

Ideas of this kind found their earliest expression in the arts. In our own time they have graduated to the extremely sophisticated debates surrounding research in artificial intelligence. But, significantly for our purposes here, in the mid-nineteenth century such themes also began to appear in the literature of political theory. The writings of Karl Marx on labor, manufacturing, and machinery contain passages which develop the theme of autonomous or, as Marx preferred, alienated technology and which employ images of technological animism. Marx believed that under the conditions of nineteenth-century capitalism, technology had taken on an independent, malevolent, lifelike existence and stood opposed to man as an alien and even monstrous force. In *Capital* Marx describes the advanced factory system of his time as "a huge automaton . . . driven by a self-acting prime mover . . . [which] executes, without man's help, all the movements requisite to elaborate raw material,"[50] words that Forster or Vonnegut might well have used. "In the factory," he notes, "we have a lifeless mechanism independent of the workman, who becomes its mere living appendage."[51] "An organised system of machines, to which motion is communicated by the transmitting mechanism from a central automaton, is the most developed form of production machinery. Here we have, in the place of the isolated machine, a mechanical monster whose body fills whole factories, and whose demon power, at first veiled under the slow and measured motions of his giant limbs, at length breaks out into the fast and furious whirl of his countless working organs."[52] Such references to a mechanical monster which makes men its appendages are not mere instances of rhetorical excess. They are an important aspect of Marx's view of the profoundly pathological relationship that men had to industrial technology, a view that was in turn an outgrowth of Marx's theory of alienated labor.

For Marx, technology must be understood within the context of man's fundamental relationship to nature. This relationship is one in which an active, inherently sensuous being confronts material reality for both sustenance and continued development. Like other animals, man depends on nature for his very survival. But unlike them, man has the capacity for free, conscious, productive *activity*, which permits him not only to survive but also to develop his potential as a member of the human species. "The practical construction of an objective world, the manipulation of inorganic nature, is the confirmation of man as a species being."[53] "Productive life is . . . species life. It is life creating life."[54] Man lives in constant contact with material reality and shapes it to his own purposes. Through his labor, he pours his life into nature and makes it part of himself. In this sense man has two bodies, the one with which he is born and an inorganic body, which is nature. Marx spoke of the creative expansion and extension of man into the world through labor as self-activity. An individual's highest realization comes in totally free self-activity in community with his fellows. In this situation man's relationship to technology is an entirely positive one. Man uses all of the instruments and tools—the productive forces—available to him in a conscious, productive manner. "The appropriation of these powers is itself nothing more than the development of the individual capacities corresponding to the material instruments of production. The appropriation of a totality of instruments of production, is for this very reason, the development of a totality of capacities in the individuals themselves."[55]

In the course of history, however, the original and proper relationship had gone wrong. Through a long series of technological and social developments culminating in the capitalist industrial order, men had fallen into a condition in which the means of life, freedom, and enrichment had evolved into forces of degradation, slavery, and death. They had become alienated from productive life, from their products, tools, labor, and their fellows. No longer was the laborer able to realize genuine human potential through self-activity. Labor had become a drudgery, which just barely provided for the necessities of life. Man the laborer continued to pour his humanity into an endless succession of products, but this process did not extend his power or satisfaction. Rather, it drained life's essence. "The more the worker expends himself in work the more powerful becomes the world of objects which he creates in face of himself, the poorer he becomes in his inner life, and the less he belongs to himself. It is just the same as in religion. The more of himself man attributes to God the less he has left in himself. The worker puts his life into the object, and his life then belongs no longer to himself but to the object. The greater his activity, therefore, the less he possesses. What is embodied in the product of his labour is no longer his own. The greater the product is, therefore, the more he is diminished."[56]

What this situation meant in E. M. Forster's story, it also signifies to Marx. The alienated transference of life produces a strange phenomenon in which man's own being is experienced as something foreign, removed, independent, and threatening. "The alienation of the worker in his product means not only that his labour becomes an object, assumes an external existence, but that it exists independently outside himself, and alien to him, and that it stands to him as an autonomous power. The life which he has given to the object sets itself against him as an alien and hostile force."[57] Under the conditions of capitalist technology men are no longer the masters of their tools, products, or productive social relationships. Neither do any of these things benefit them. Quite the contrary, the workers are worse off than before. The coming of the technical marvels of the industrial revolution has driven them to the lowest point of human existence. "By means of its conversion into an automaton, the instrument of labour confronts the labourer, during the labour process, in the shape of capital, of dead labour, that dominates, and pumps dry, living labour power."[58] Describing this state of affairs, Marx formulates what amounts to the first coherent theory of autonomous technology. His words in *The German Ideology* effectively summarize much of what is problematic for us in this inquiry: "This crystallization of social activity, this consolidation of what we ourselves produce into an objective power above us, growing out of our control, thwarting our expectations, bringing to naught our calculations, is one of the chief factors in historical development up till now."[59]

Marx believed that man could achieve free and productive social existence only by overcoming the alienation of labor. This step would, of course, come through a revolution and the advent of a communist social order, an important aim of which would be to liberate technology for its truly human function and to make it impossible that pathological formations of the past could ever be reestablished. "The reality, which communism is creating, is precisely the real basis for rendering it impossible that anything should exist independently of individuals, in so far as things are only a product of preceding intercourse of individuals themselves."[60] "Only at this stage does self-activity coincide with material life, which corresponds to the development of individuals into complete individuals and the casting off of all natural limitations."[61] When this is achieved the mystification of autonomous technology will vanish. No longer will men be confronted with an "alien and hostile force" made up of their own products. "The life-process of society, i.e., the process of material production, will not shed its mystical veil until it becomes the product of freely associated men, and is consciously regulated by them in accordance with a settled plan."[62]

But one should not go too far in attributing a theory of autonomous technology to Karl Marx. Although he presents a well-developed conception of technology out of control, in the context of his work as a whole it is merely an interlude in the midst of a much larger argument.[63] For Marx, the alienation of labor and the appearance of a massive, life-draining industrial mechanism cannot be considered independently of the historical class struggle, surplus value in industrial economics, capitalist accumulation, and the social and political domination of the bourgeoisie. Marx introduces the specter of an alien technological force to illustrate how far historical alienation has gone and how little power the worker has over his own daily existence. But Marx makes it clear that mastery over technology has not been really lost. It has simply been removed to a small segment of the social order, the capitalist class. Directly following the animistic passages of *Capital* quoted above, he stresses that there *is* most definitely a "master" "in whose brain the machinery and his monopoly of it are inseparably united."[64] The individual worker is indeed dwarfed and diminished by his relationship to the technical means with which he has contact. But behind the mechanism stands a human figure who takes up and manipulates for his own advantage the power that has been drained from the proletariat. "The special skill of each individual insignificant factory operative vanishes as an infinitesimal quantity before the science, the gigantic physical forces, and the mass of labour that are embodied in the factory mechanism and, together with that mechanism, constitute the power of the 'master.' "[65] Again, this conclusion comes directly from the elaboration of Marx's theory of alienated labor. In the first of the *Economic and Philosophic Manuscripts* he insists that technology personified and animated can mean only that there is a real person or class of persons lurking behind the strange phenomenon. "The *alien* being to whom labour and the product of labour belong, to whose service labour is devoted, and to whose enjoyment the product of labour goes, can only be *man* himself. If the product of labour does not belong to the worker, but confronts him as an alien power, this can only be because it belongs to a *man other than the worker.*"[66]

But is the role of the master any longer necessary? Does the mechanism actually require his presence and guiding hand? On this question, twentieth-century notions of autonomous technology diverge from those in *Capital.* In the writings of Ellul and Marcuse, which in many ways follow Marx's analysis, the willful, exploiting subject is no longer crucial to the operation of the system or to its domination of mankind.

The master is in a true sense a redundancy, and his governance is ornamental rather than decisive. The privileged position of an elite or ruling class is not proof that it steers the mechanism but only that it has a comfortable seat for the ride. Ultimately the steering is inherent in the functioning of socially organized technology itself such that *any* elite, class, or ruling body "at the helm" would be forced to follow its necessary course. Here we catch a first glimpse of one of the interesting political problems posed in theories of autonomous technology: if Poe were doing his investigation today, he would probably not find a man hidden in the "Chess Player"; a human being would only gum up the works.

A fascinating, sprawling masterwork in the literature of autonomous technology in our time is Ellul's *La Technique ou l'enjeu du siècle* (literally, "Technique: The Stake of This Century"), translated as *The Technological Society*.[67] Inheritor of the tradition I have sketched, the book makes themes long recognized in fiction, poetry, film, and the plastic arts accessible to contemporary sociology and political theory. For reasons having to do with different intellectual climates and states of technical development, Ellul's thinking has attracted more attention in the United States than in his native France. The work appeared in English at a time when cybernetics, computers, and systems analysis were becoming subjects for widespread academic and public debate. In its very outrageousness, Ellul's argument challenged American readers and cast a fruitful light on a variety of their inchoate perceptions.

More than any other single work, *The Technological Society* is the starting point for the questions I will be considering here. While it is scarcely the last word on the subject and while its formulations are idiosyncratic and often flawed, Ellul's book is true to its original French title; it does provide a vivid idea of how technology is central to what is at stake in this century. Ellul is often criticized for reification and anthropomorphism in his major concepts. Indeed, his writing is filled with constructions of the sort: "technique pursues its own course"; "technique advocates the entire remaking of life"; and "technique tolerates no judgment from without." The charge of reification, however, loses some of its impact if one considers that social science consistently reifies such concepts as "society," "family," and "bureaucracy." One is hard pressed to think how it could do otherwise. Since we cannot have all that we wish to talk about immediately present as empirical referents, we must employ symbols to represent phenomena. Ellul employs "technique," a concept we are not accustomed to using in a reified manner, to speak of technical things in general.

The claim that "technique" becomes anthropomorphic in his writing is more to the point, precisely to the point in fact. Ellul's main thesis, after all, one that he takes full responsibility for and works to clarify with argument and example, is that technique is entirely anthropomorphic because human beings have become thoroughly technomorphic. Man has invested his life in a mass of methods, techniques, machines, rational-productive organizations, and networks. They are his vitality. He is theirs. In body, mind, will, and activity they must now move in unison or both will perish. By employing the metaphor of technological animism—referring to "technique" as a sensing, thinking, deciding, demanding subject—Ellul offers us an image that encompasses not only the substance of his own complex arguments but many similar conjectures and hypotheses in Western literature of the past century and a half.

My purpose from this point will be to examine a series of specific topics within the general theme of autonomous technology. I will be asking: When have reports that "technology is out of control" actually come up? Which experiences in modern life lie behind such conceptions? Some of the work, therefore, will consist of sorting, classifying, and analyzing a variety of observations or hypotheses from sources in philosophy, social science, and everyday life. At the same time we will be engaged in a continuing process of criticism, for one must always ask: What sense do these notions make? Are they in any way valid? My method here will be to seek the rational core of what are sometimes less than rational positions, making it clear what those positions take into account and what their meaning for us might be.

Our path begins in the next chapter with a consideration of several issues centering on the phenomenon of technological change. In a brief chapter following, I consider attempts made by a number of schools of thought to locate the ultimate origins of Western technological dynamism. Turning to politics in chapter 4, I will examine the major tenets, historical and contemporary, of theories of technocracy. Chapters 5 and 6 will be devoted to an elaboration of the concepts and major assertions of a theory of technological politics. This view goes beyond an exclusive concern for technocratic elites to focus upon the role of technical structures and processes in modern culture as a whole and moves from there to a redefinition of the terms of political life. A separate chapter will then weigh the special problem of responsibility as it confronts institutions of increasing scale and technical complexity. For me the subject of responsibility represents the absolute minimum level at which anyone might become practically involved in the substance of the intellectual topics within our scope. A concluding chapter offers thoughts for those who wish to go further than this bare minimum to reevaluate the role of technics in their lives and to seek new horizons for a politics of technology.

The reader is justified in wondering: What is the writer's own position on this matter? Does he actually maintain that technology is out of control? Is that the point? A reasonable answer, I would respond, can only be given with reference to any one or more of the specific subthemes I shall outline. Some of the dilemmas described are more crucial, some of the problem formulations more substantial than others. My intention is to demystify the issues rather than to dismiss or debunk them. From there the reader is asked to decide what makes sense and to form his or her own judgments. Autonomous technology is ultimately nothing more or less than the question of human autonomy held up to a different light. And those who remain supremely confident about our prospects there have not been paying attention to what is happening everywhere about them.

Chapter 2
Engines of Change

Following his visit to the Great Exposition in Paris in 1900, Henry Adams, the American historian and philosopher, reported a peculiar fascination with one part of the exhibit—the hall of dynamos. In his *Education*, Adams explains that his friend and guide through the exposition, Samuel Langley, was able to see the machines in their most mundane aspect. "To him the dynamo was but an ingenious channel for conveying the heat latent in a few tons of poor coal hidden in a dirty engine house carefully out of sight." But to Adams, "the dynamo became a symbol of infinity."[1]

> As he grew accustomed to the great gallery of machines, he began to feel the forty-foot dynamos as a moral force, much as the early Christians felt the Cross. The planet itself seemed less impressive, in its old-fashioned, deliberate, annual or daily revolution, than this huge wheel, revolving within arm's length at some vertiginous speed. . . . Before the end, one began to pray to it; inherited instinct taught the natural expression of man before silent and infinite force. Among the thousand symbols of ultimate energy, the dynamo was not so human as some, but it was the most expressive.[2]

What Adams saw that day became for him the prime representation of a set of perceptions that many before him and since have shared. Behind the tremendous power of the giant machines stood the forces of natural energy that man in increasing measure was able to harness for his own purposes. These forces were in turn linked to a new force in human history—an ever-accelerating change in the conditions of social, economic, and political life. Convinced that this rate of change and the momentum it brought to the affairs of man were the most important facts of his time, Adams set about to discover laws of history to account for the ever-expanding flow of technological energy. The dynamo was symbolic of a process so colossal and staggering to the imagination that only a mathematical series or the idea of divinity could help the historian fathom its nature and probable course. Adams's sense of wonder was matched only by his intuition of danger. As he wrote in a letter to his brother Charles, "You may think all this nonsense, but I tell you these are great times. Man has mounted science and is now run away with. I firmly believe that before many centuries more, science will be the master of man. The engines he will have invented will be beyond his strength to control."[3]

Since Adams's time, the type of machine used to symbolize the course of technological development has shifted with the fashions of the moment. The automobile, the airplane, the nuclear reactor, the space rocket, the computer—all have stood as representations of the now familiar set of phenomena: the growth of scientific knowledge, the expansion of technics, and the advent of rapid social change. In its turn each new piece of miraculous apparatus has been heralded as the essence of a new (but usually short-lived) "age" in the history of mankind. In its turn each machine metaphor has opened areas of both insight and radical blindness as it becomes a means of interpreting what happens in our world.[4] Beyond the diversity of metaphors, however, lies a fundamental shared perception that modern history is characterized by a process of continuing change and that somehow machines and other manifestations of new technology are at the center of this process.

One way of answering the question, Why is technology problematic? rests on exactly this point. Technology is a source of concern because it changes in itself and because its development generates other kinds of changes in its wake. For many observers this is the whole story, the alpha and omega of the entire subject. To look for crucial questions is to look for inventions, innovations, and the myriad of ramifications that follow from technological change.

It will eventually become clear why I do not see the matter from this perspective alone. Change is merely one category, by no means the most interesting one, within the range of problems I want to examine. Nonetheless, among the significant themes included in the notion of autonomous technology are those pointing to the process of technological development and historical change as something that eludes control. The matter of understanding precisely how various specific changes in technology have affected the course of social change must be left to the historian. My concern here will center on the questions raised in the previous chapter: human autonomy and the loss of mastery. Lewis Mumford began his career by inquiring into purely factual aspects of technological change and ended with the question, What does technology have to do with *not* being free? It seems to me that this, rather than purely empirical or historical topics, is the urgent subject at present.

Momentum and Motive

Let us begin by noticing a curious paradox that plagues almost all discussions of technological change, both our ordinary common-sense view of things and more precise scholarly conceptions. On the one hand we encounter the idea that technological development goes forward virtually of its own inertia, resists any limitation, and has the character of a self-propelling, self-sustaining, ineluctable flow. On the other hand are arguments to the effect that human beings have full, conscious choice in the matter and that they are responsible for choices made at each step in the sequence of change. The irony is that both points of view are entertained simultaneously with little awareness of the contradiction such beliefs contain. There is even a certain pride taken in embracing both positions within a single ideology of technological change.

Signs of the paradox emerge from the ways in which historical developments of the past two centuries are normally represented. For several generations, it has been commonplace to see technological advance in the context of a vast, world-transforming process—industrialization, mechanization, rationalization, modernization, growth, or "progress." "Industrialization," until recently the most popular label, points to the range of adaptations in social, technical, and economic structure that societies have undergone in order to support the large-scale production of material goods. A more fashionable term at present, "modernization," attempts to correct for the narrowness of the industrial concept in light of twentieth-century history. In essence it means "all of those changes that distinguish the modern world from traditional societies." During the last two or three centuries there has been an astounding increase in the scope, variety, sophistication, and effectiveness of man's scientific and technological activity. As the knowledge of physical reality has expanded, men have been able to exploit immense new sources of energy and materials and to devise larger, more complex, more productive forms of manufacturing, agriculture, transportation, communication, medicine, and warfare. Along with these developments have come a vast array of social, economic, demographic, and political changes which bring a whole new character to civilized life—increased per-capita income, longer life expectancy, rapid expansion of world population, rise in literacy, proliferation of social roles, and so on.[5]

In most writings on "industrialization," "modernization," and "development," the element of technological change is found to be so crucial to the origins and continued formation of everything modern that it is included in the definition of the process under study. W. W. Rostow speaks of technological constraints that existed in "pre-Newtonian" times and asserts that when modern technology was lacking, social change was lacking as well. "Ultimately," he observes, "these constraints, operating in complex ways, were judged the cause of the cyclical patterns typical of the history of traditional societies."[6] But with the coming of Newtonian science and the industrial revolution, a pattern of linear growth was established that continues to the present day. In response to the developments in science and technics, the institutions of society and politics alter their structure in order to "absorb" the new technologies. "Political development," Rostow tells us, "consists in the elaboration of new and more complex forms of politics and government as societies restructure themselves so as to absorb progressively the stock and flow of modern technology."[7] In Wilbert E. Moore's book *Social Change,* the point is even more definite: "What is involved in modernization is a 'total' transformation of a traditional or pre-modern society into the types of technology and associated social organization that characterize the 'advanced,' economically prosperous, and relatively politically stable nations of the Western World."[8]

Beginning in the nineteenth century and continuing to the present day, there has been a significant number of writers ready to claim that such historical tendencies reflect a process of technological change that is self-generating, self-determining, and in a true sense inevitable. For our time, Jacques Ellul has offered a remarkably pure version of this position. His writings cite an endless barrage of examples which attempt to demonstrate that "technique has become a reality in itself, self-sufficient, with its own special laws and its own determinations."[9] Others have tried to identify precisely what the special laws are that govern the dynamics of technological transformation. For writers like Roderick Seidenberg, Leslie White, and their notable precursor, Henry Adams, such changes are best seen as a purely physical process. Alterations in the material culture of modernity are, in this view, much like those that occur in the motions and changes of state of inanimate objects. It makes sense, therefore, to speak of them in terms of speed, force, momentum, energy, and acceleration and to account for the aggregate process under laws of strict, causal determinism.

Adams, for example, believed he had discovered a "law of acceleration" in human history based on the fact that the complexity of civilization, as well as its control over physical force, seemed to double and redouble in an ever-increasing ratio. Adams even suggested methods of measuring the acceleration: "The coal output of the world, speaking roughly, doubled every ten years between 1840 and 1900, in the form of utilized power, for the ton of coal yielded three or four times as much power in 1900 as in 1840. Rapid as this rate of acceleration seems, it may be tested in a thousand ways without greatly reducing it."[10] Such a process of increasing energy and organization would, Adams believed, defy all attempts at conscious direction or opposition. "A law of acceleration, definite and constant as any law of mechanics, cannot be supposed to relax its energy to suit the convenience of man."[11]

For Adams and his present-day followers the specific physical laws that govern historical development are those of thermodynamics, the science of the relationship of heat and mechanical energy. Restated by

Seidenberg, the fundamental "discovery" of this school is "that the stages in the course of human evolution may be comparable to changes of state in a purely material system as expressed by the Rule of Phase in thermodynamic theory."[12] Such changes in "phase" involving an increase in energy and organized complexity are seen as an anti-entropic movement in physical reality. The movement runs directly counter to the second law of thermodynamics, which states that entropy increases, that is, matter becomes increasingly disorganized and energy increasingly diffused. In what certainly must be the ultimate in organization theories, the acceleration of technological and social development is depicted as a final, futile fizzle in a universe headed for thermodynamic annihilation.

Adams's law of acceleration is strikingly similar to recent assertions about the problem of "exponential growth" and is subject to many of the same objections. In trying to account for the skyrocketing use of coal, Adams would have done better to cite such reasons as the expansion of America's population, the increase in number of industrial cities, and the shrinking of the western frontier. Historical analogies to scientific laws usually go adrift by making claims at the wrong level of explanation.

But are Ellul and the thermo-organization theorists entirely unique in attributing a certain dynamism and ineluctability to the process of technology-associated change? The scholarly literature on industrialization, modernization, and growth contains continued reference to such concepts as speed, momentum, force, and acceleration. The language of change in this area is inextricably bound to such conceptions as "the dynamics of modernization," "the forward thrust of technological change," "the accelerated pace of industrial development." The most famous of the recent metaphorical devices here is W. W. Rostow's airplane image, "the take-off and drive to technological maturity." The attribution of an inherent dynamism to the process of change is particularly evident in the use of "-ization" suffix words. Here, perhaps unintentionally, connotations of a self-generating, self-sustaining process frequently creep in. There is a tendency to speak as if industrialization, modernization, and the other "-izations" are similar to such physical processes as ionization in the sense that once underway, they continue on their own with a kind of in-built necessity or inertia. To a certain extent, this connotation is merely a quirk in our language. In English, the "-ize" verbs do imply an active, willing subject, while the "-ization" nouns suggest a self-generating process. In many instances, however, economists and historians wish to maintain that the "-ization" they are studying is in fact a phenomenon *sui generis*, inertial and self-sustaining. In David Landes's excellent work on technological change, *The Unbound Prometheus*, we find statements of the following sort: "It was the Industrial Revolution that initiated a cumulative, self-sustaining advance in technology whose repercussions would be felt in all aspects of economic life."[13] This, of course, is similar to Rostow's major point. After the "takeoff," industrial societies enter a period of sustained development in which growth feeds upon growth in an ever-expanding, ever-accelerating upward curve.

But there is more here than just the convenience of a metaphor. In describing the course of technical and social history, scholars often commit themselves to a view that finds dynamism and inertial change apparent in certain objective facts. C. E. Black finds the impact of modernization to be a universal phenomenon. "The process of adaptation," he observes, "had its origins and initial influence in the societies of Western Europe, but in the nineteenth and twentieth centuries these changes have been extended to all other societies and have resulted in a world wide transformation affecting all human relationhsips."[14] The process is commonly held to be irreversible in the sense that the changes have taken us so far from the forms of traditional society that it would be impossible ever to move in that direction again. This concept is made abundantly clear in Clark Kerr's *Industrialism and Industrial Man*. According to Kerr, "An argument against industrialization in general is now futile, for the world has firmly set its face toward the industrial society, and there is no turning back."[15] The changes, similarly, are held to be inevitable in the special sense that new forms of technology and social life must necessarily replace older modes. The technics and social configurations of former times are doomed and simply cannot survive. "Once under way," Kerr observes, "the logic of industrialization sets in motion many trends which do more or less violence to the traditional pre-industrial society."[16] The only alternative is to yield to these trends and obey their imperatives. Witness the fate of the reluctant industrial workers: "The dynamic science and technology of the industrial society creates frequent changes in the skills, responsibilities, and occupations of the work force. . . . The work force is confronted with repeated object lessons of the general futility of fighting these changes, and comes to be reconciled, by and large, to repeated changes in ways of earning a living."[17]

Conclusions of this sort are usually offered as purely factual matters. The historians, economists, and political scientists who tell us of such things are merely reporting an objective state of affairs that we can verify for ourselves. At the same time there is, as Henry Adams would have expected, a tendency to adopt a worshipful attitude toward the "-izations" and their juggernaut-like advance. One begins to see the process as an overwhelmingly powerful destiny with the moral obligations of service and obedience. In this light David Apter compares the modern condition to the plight of Sisyphus in Camus' fable: "Sisyphus, returning again and again to roll his rock up the hill, may appear absurd. Yet on each occasion he is happy. How odd that seems: And how like our own times. The work of modernization is the burden of this age. It is our rock."[18] Another student of modernization, Myron Weiner, is even more definite in this regard. "Social scientists," he asserts, "are concerned not only with how modernization takes place, but also with how it can be accelerated."[19] Faced with a world undergoing many comprehensive changes, the social scientist must ask himself, "What are these changes, how are they related, how do we study them, how can these changes be hastened?"[20] The objective, neutral disposition of social science apparently vanishes when it encounters the inevitable. One must affirm it and lend a hand. It is, after all, our rock.

With the articulation of such themes the distinction between orthodox scholarship and the literature of autonomous technology begins to blur. Arrayed together in a list, these attributes of modernization—dynamism, universality, pervasiveness, irreversibility, inevitability, and positive destiny—cover most of what Ellul includes in his account of the proliferation of *la technique*. While Ellul's treatment goes much further than normal social science in seeing the development as monistic and self-augmenting, the general implication of the more prudent studies is not that much different. The modern age is portrayed as a time in which individuals, groups, nations, and whole cultures are caught up in a process of technological transformation which affects every sphere of life and which pulverizes preexisting social alternatives.

But, surely, we feel there must be some important differences between the views of critics of technological society and those of scholars of modernization. At the very least, the two points of view give different impressions when we read them.

One obvious issue separating the two perspectives—and, lamentably, the one most frequently commented upon—is whether the direction of change has been a blessing or curse for mankind. As noted above, most students of development and modernization hold that the process is undeniably beneficent. While they admit that technological advance has frequently brought disaster to persons caught between the old ways and the new, they consider these costs in human wreckage to be unavoidable and, on the whole, worthwhile.[21] The general condition of mankind has improved immeasurably. In contrast Ellul and other critics of technology question the good of the specific kinds of changes that have occurred and argue that human freedom, dignity, and well-being have not been enhanced by the historical flood tide.

While much has been made of the gap between optimism and pessimism and whether a particular writer is a "prophet of hope" or a "prophet of doom," this is, I think, in the end a vacuous distinction. In much scholarship today there exists an almost compulsive need for optimism on this topic. If one notices that an Ellul or Mumford or some other author is pessimistic in his conclusions, that becomes sufficient ground for dismissing anything he or she might be saying. Pessimism, it is argued, leads to inaction, which merely reinforces the status quo.[22] This is somehow different from optimism, which leads to activity within the existing arrangement of things and reinforces the status quo. In the same manner of thinking one would have to do away with Aeschylus, Shakespeare, Melville, and Freud and much of what is deepest and most illuminating in Western literature.

But setting aside the invidious matter of gloom versus hopefulness, we encounter an issue of genuine interest. Are the various paths of technological development freely and deliberately chosen, or are they instead the product of determinism, necessity, drift, or some other historical mechanism? To what extent are man's active intelligence, moral and political agency, and capacity for control true determinants of technical and social progress? On the face of the matter it appears that voices in the discussion about technological change can easily be sorted into two basic positions on this theme. One side affirms freedom and the reality of "choice," while the other sees mankind as a pawn in the indelicate hands of history. Or so it seems.

Under the voluntarist way of looking at things the notion that people have lost any of their ability to make choices or exercise control over the course of technological change is unthinkable; for behind the massive process of transformation one always finds a realm of human motives and conscious decisions in which actors at various levels determine which kinds of apparatus, technique, and organization are going to be developed and applied. Behind modernization are always the modernizers; behind industrialization, the industrialists. Science and technology do not grow of their own momentum but advance through the work of dedicated, hard-working, creative individuals who follow highly idiosyncratic paths to their discoveries, inventions, and productive innovations. In the process of development, active, thinking agents—James Watt, Thomas Newcomen, Thomas Watson, Alfred Sloan, the Du Ponts—are clearly present at each step, following distinctly human ideas and interests. Societies, furthermore, do not yield passively to the "thrust" of modernization. Political and economic actors of the world's nation-states make conscious decisions about what kinds of technological development to encourage and then carry out these decisions in investments, laws, sanctions, subsidies, and so on. In some instances—nineteenth-century China, for example—the introduction of modern technology was actively opposed. In such cases "development" could not begin until a Western colonial power had neutralized such opposition in a colonized country or until an internal political upheaval had put men favorable to such changes in positions of leadership. The modern history of technological change is, therefore, not one of uniform growth. It is instead a diverse collection of patterns rooted in specific choices that individuals, groups, and nations have made for themselves and imposed on others.[23]

The voluntarist position on such questions sounds eminently reasonable having, as it does, the clear timbre of common sense. Yet there is a difficulty here that is not easily dismissed, a difficulty that carries us to the paradox mentioned earlier. In much of the literature on technological change two radically different languages are spoken at once, with little attention to how they can both describe the same set of events. One is the language of a dynamic global process that moves ineluctably forward, transforming everything in its path. The other is the language of free agency, individual will, deliberation, and choice in which the path of technological advance is consciously directed. Facing this paradox, it does little good to argue that the "-izations" are merely conveniences of speech, shorthand ways of talking about myriads of free decisions and actions. We have already seen that writers on industrialization, modernization, and the like *do* wish to assert that the process itself has a true momentum or inertial beyond individual or even collective will. Thus, even a judicious spokesman on industrial development, Clark Kerr, is able to exclaim:

> This technique knew no geographical limits; recognized no elites or ideologies. Once unleashed on the world, the new technique kept spreading and kept advancing.
>
> It is the great transformation—successful, all-embracing, irreversible.
>
> By the middle of the twenty-first century industrialization will have

swept away most pre-industrial forms of society, except possibly a few odd backwaters.[24]

To escape the dilemma here, scholars often resort to the view that human freedom actually exists within the limits set by the historical process. While not everything is possible, there is much that can still be chosen. This perspective enables David Apter to see modernization as "a process of increasing complexity in human affairs within which the polity *must act* [emphasis added]" and at the same time to hold that "to be modern means to see life as alternatives, preferences and choices."[25] Rostow, in the same vein, sees the process of technological development as a grand staging ground, which gives shape to all of society's most important decisions. "With the take-off and drive to technological maturity the process of industrialization itself becomes the center of politics."[26] "The efficient absorption of technologies," he notes, "carries with it powerful imperatives, social and political as well as economic."[27] The effect of these developments is to lay on "the agenda a succession of pressures to allocate the outputs of government in new ways."[28] But once the fundamental agenda has been set, there is considerable choice about the specific sociotechnical forms the development will take.

To Ellul, Marcuse, Mumford, and other critics of the technological society, arguments of this stripe are entirely in vain. The self-confidence of the modernizers is merely a guise concealing a strict obedience to the momentum of events. Under present conditions men are not at all the masters of technological change; they are its prisoners. Although the voluntarists may celebrate man's shrewdness and freedom, the celebration cannot alter the condition that their own theories reveal. The shout of freedom, D. H. Lawrence noted long ago, "is the rattling of chains, always was."

The critics themselves, however, do not escape the paradox but approach it in a different manner. Mumford, Marcuse, Ellul, and Goodman never deny human freedom as a metaphysical or historical possibility. What they do assert in various ways is that with regard to technological change, such possibilities remain unrealized. "The reader must always keep in mind," Ellul cautions his American audience, "the implicit presupposition that *if* man does not pull himself together and assert himself (of if some other unpredictable but decisive phenomenon does not intervene), then things will go the way I describe."[29] The process of technological advance surges along an ineluctable path largely because human agents have abdicated their essential role. Thus, if the development of technics in civilization does not proceed by a strict cause-and-effect chain, it might as well be doing so.

The contradictions and tensions we see here are not the exclusive property of philosophers and social theorists. They are, I would argue, crucial to our common sense views of the matter as well. Journalism, popular literature, and subway conversations frequently exhibit the puzzle that while someone, somewhere is making decisions and choices, things are also running amok. Autonomous technology in the sense of technological-change-out-of-control is a part of the modern experience. Our task is to see if such experience can be described in an intelligible way. There are two areas, I believe, where it pays us to look.

First, are there certain aspects of the process of technical change that do not depend on the elements of free will, conscious decision, or anyone's intelligent control? Are there some important points at which there simply is no effective, guiding human will that determines the final result?

Second, is it the case, as the critics have argued, that the capacity for modern men to make free choices to counter the anarchic tendencies of technological development has itself been eroded by an obsessive attachment to the mode of technological change? Obviously, in some sense "choice" is always possible. But is that enough? An odd freedom it is that in every instance is capable only of choosing the next step in the logic of technical expansion.

In the sections that follow I shall draw upon a variety of cases and arguments that contain the substance of a theory of autonomous technological change. Far from being expressions of mere pessimism or anxiety, the themes we will consider—technological evolution, determinism, drift, and the technological imperative—point to some genuine dilemmas in the ways in which modern societies undergo change, as well as in our conceptions about such change.

Technological Evolution

A central theme in the speculative writing on technology of the past century is that forms of technics, like forms of biological life, undergo a process of evolution. With the passage of time, newer and more sophisticated varieties of apparatus, organization, and technique rise to replace older, simpler varieties. New technologies enter into areas of social existence where they had not been previously. Just as Darwin observed that the various species of life on the Galapagos Islands tended to specialize and diversify into particular biological niches, so it is that forms of technology continually spread into fresh areas of social utility. In both number and diversity the kinds of technical artifice available to human societies increase. One of the first thinkers to notice the similarity between biological and technical evolution was Karl Marx. In *Capital* he comments: "Darwin has aroused our interest in the history of natural technology, i. e., in the formation of the organs of plants and animals, as instruments of production for sustaining life. Does not the history of the productive organs of man, of organs that are the material basis of all social organization deserve equal attention? And would not such a history be easier to compile, since, as Vico says, human history differs from natural history in this respect, that we have made the former, but not the latter?"[30] Marx, as we shall see later, definitely did not believe that the evolution of technology was a self-generating process. But there have been many who have entertained this notion, either as an interesting literary device or a serious proposition.

As an exercise in thinking, of course, it is possible to look upon technological change as a kind of evolutionary flow. Characteristic of such applications of Darwinian theory is a tendency to reduce the human role to a distinctly secondary status. Since the theory focuses upon the evolving forms of technics in themselves, human beings come to be seen as the mere carriers of technology. Each generation bears and extends the technical ensemble and passes it on to the next generation. The mortality of human beings matters little, for technology is itself the immortal and, therefore, the more significant part of the process. Specific varieties of technics can be compared to biological species that live on even though individual members of the species perish. Mankind serves a function similar to that of natural selection in Darwinian theory. Existing structures in nature and the technical ensemble are the equivalent of the gene pool of a biological species. Human beings act not so much as participants as a selective environment which combines and recombines these structures to produce new mutations, which are then adapted to a particular niche in that environment.

Particularly among thinkers with backgrounds in the natural sciences, engineering, and anthropology, this perspective is occasionally suggested as a valid portrait of the development of technology. The fact that human beings—both as individuals and as a species—are dwarfed and placed in a secondary status by these developments is seldom taken as a warrant for gloom. Ellul, of course, draws some rather bleak conclusions from the Darwinian strand in his thought: "Technique pursues its own course more and more independently of man. This means that man participates less and less actively in technical creation, which, by the automatic combination of prior elements, becomes a kind of fate. Man is reduced to the level of a catalyst."[31] "At the present time, technique has arrived at such a point in its evolution that it is being transformed and is progressing almost without decisive intervention by man."[32] "It is evolving with a rapidity disconcerting not only to the man in the street, but to the technician himself."[33]

But there are other thinkers who hold that the evolution of technology is the most glorious of blessings, even though it spells the obsolescence of the human race. The movement is leading to a higher stage of development in the history of the world and in the history of consciousness. Support for this hypothesis is found in what its proponents believe to be the convergence of men and machines. Machines, computers in particular, are becoming more and more human and are, according to some estimates, on the verge of a consciousness and intelligence of their own. Human beings, on the other hand, are becoming more and more like the cyborgs—cybernated organisms—of science fiction. They now require an increasing number of technical support systems to keep them running efficiently: cardiac pacemakers, artificial kidneys, computer-aided systems of inquiry, and so on. Eventually there will be a kind of total man-machine symbiosis in which the organic parts of the human being will be grafted directly to highly sophisticated, miniaturized technological organs, which will assist in all physical and intellectual functions. The world will enter a stage in the development of consciousness and physical performance much advanced over anything in our experience.[34]

One cost of this remarkable set of advances, according to the technoevolutionists, may well be that the fleshly part of the new technological ensemble will become a drag on an otherwise highly efficient system. In the words of Arthur Clarke, "Can the synthesis of men and machine ever be stable, or will the purely organic component become such a hindrance that it has to be discarded? If this eventually happens—and I have . . . good reasons for thinking that it must—we have nothing to regret and certainly nothing to fear." "The Tool we have invented is our successor. Biological evolution has given way to a far more rapid process—technological evolution." "No individual exists forever," he continues: "why should we expect our species to be immortal? Man, said Nietzsche, is a rope stretched between the animal and the superhuman—a rope across the abyss. That will be a noble purpose to have served."[35]

In other words, man should be pleased to have played even a small walk-on part in this much larger drama. To complain that humans have been left out of the final scenes is merely an example of outdated species chauvinism. But whether taken in a positive or negative light, all theories of technoevolution suffer from the same basic flaw. Their major discovery—the eclipse of mankind—turns out to be something they had assumed in the first place. The evolutionary perspective begins with the adoption of abstract categories which do not include a role for free, conscious human agents. Is it any wonder that when theorists of this school go looking for man, they find that the human role has shrunk to virtually nothing?

There is, however, a more concrete level at which a variant on the idea of technoevolution has considerable power. If we examine the progress of scientific discovery and technological invention and innovation in modern history, we do encounter something resembling an evolution of forms. With the passage of time both science and technics become more specialized, diversified and complex. It is this aspect of the process of change which Ellul labels the "self-augmentation of technique," and which he sees as one of the origins of the modern malady.

Ellul's discussion begins by pointing out that modern technical progress usually comes as the result of organized social effort. Modern men are thoroughly convinced of the inherent superiority of technique. In every trade and profession they work in combination to devise technical improvements. "Technical progress and common human effort," Ellul observes, "come to the same thing."[36] It is important to note here that Ellul holds that there is no longer any real distinction between science and technique.[37] Indeed, he states that what is usually called "science" has become so thoroughly technicized that it is now "an instrument of technique." By "technical progress," therefore, Ellul includes the discoveries made within the social system of science along with the more mundane advances in other fields of knowledge and practice.

With this view stated, Ellul seeks to deny the element of voluntarism implied by the idea of "common human effort." What is important, he asserts, is not the discovery or invention of any particular individual or group but, instead, "the anonymous accretion of conditions for the leap ahead."[38] "When all the conditions concur, only minimal human intervention is needed to produce important advances. It might almost be maintained that, at this stage of evolution of a technical problem, whoever attacked the problem would find the solution."[39] A consequence of this state of affairs is that the role of genius in scientific and technical progress has shrunk to insignificance. Those who continue to talk about it are merely propagating an illusion. According to Ellul, "It is no longer the vision of a Newton which is decisive."[40] "The accretion of manifold minute details, all tending to perfect the ensemble, is much more decisive than the intervention of the individual who assembles the new data, adds some element which transforms the situation, and thus gives birth to a machine or to some spectacular system that will bear his name."[41]

Another important aspect of self-augmentation is the continuing increase in the scope and number of techniques. "There is," Ellul maintains, "an automatic growth (that is, a growth which is not calculated, desired, or chosen) of everything which concerns technique."[42] "Apparently this is a self-generating process: technique engenders itself."[43] What is involved here is the fact that technical advances in different spheres are often closely related at their points of origin. "A technical discovery has repercussions in several branches of technique and not merely in one."[44] "When a new technical form appears, it makes possible and conditions a number of others."[45] Ellul cites the simple example of the internal combustion engine, which "conditioned" the development of the automobile, the submarine, and other devices.

In the ceaseless combination and recombination of technical advances, Ellul finds the following "law": *"Technical progress tends to act, not according to an arithmetic, but according to a geometric progression."*[46] Once again, his argument leads him to a denial of voluntarism. "What is it that determines this progression today? We can no longer argue that it is an economic or a social condition, or education, or any other human factor. Essentially, the preceding technical situation alone is determinative. When a given technical discovery occurs, it has followed almost of necessity certain other discoveries. Human intervention in this succession appears only as an incidental cause, and no one man can do this by himself. But anyone who is sufficiently up-to-date technically can make a valid discovery which rationally follows its predecessors and rationally heralds what is to follow."[47]

In a manner typical of his way of thinking, Ellul discovers an ironic determinism in the modern history of technology, a determinism heavily laden with moral quandary, for men voluntarily enter and submit themselves to social processes that generate a pattern of technical advance, which, in the end, cannot be distinguished from an ever-multiplying cause-and-effect progression. This movement is truly self-determining in the sense that its direction is inherent in the structures of technique and nature available at each stage. It is self-generating in the sense that all human motives, decisions, creative insights, and acts are placed at its service. There is nothing from the human sphere that intervenes to guide or limit its advance.

It is important to notice exactly what Ellul is saying here and on what grounds he stakes his case, for he is often "refuted" with points that do not speak to his actual position. Ellul is *not* saying that there are no choices or alternatives with regard to what is or can be done in scientific or technical development. He is not saying, for example, that there is no "science policy" in the United States to determine which kinds of inquiry are encouraged and funded and which are not. What he does say is that, given the whole or, as he puts it, the "ensemble of techniques," such specific choices by groups or individuals are insignificant. Speaking of the law of geometric progression he notes, "In arguing thus, the qualification must be made that this can be said only of the *ensemble* of techniques, of the technical phenomenon, and not of any particular technique. For every technique taken by itself there apparently exist barriers that act to impede further progress. . . . For the technical phenomenon in its ensemble, however, a limitless progress is open."[48] Ellul makes it clear that his argument rests on the idea that the "ensemble" is a distinctive entity in itself: "A whole new kind of spontaneous action is taking place here, and we know neither its laws nor its ends. In this sense it is possible to speak of the 'reality'of technique—with its own substance, its own particular mode of being, and a life independent of our power of decision."[49]

In taking this position, Ellul accepts the major premise of the tradition of sociology founded by Emile Durkheim. According to Durkheim's *Rules of Sociological Method,* "Society is not a mere sum of individuals. Rather, the system formed by their association represents a specific reality which has its own characteristics."[50] In Ellul's foreword to the American edition of *The Technological Society*, he explains (without mentioning Durkheim) the role of this notion in his work: "To me the sociological does not consist of the addition and combination of individual actions. I believe that there is a collective reality, which is independent of the individuals."[51] He also calls attention to the way in which this view has influenced his position on the issues of determinism and voluntarism. "I do not deny the existence of individual action or of some inner sphere of freedom. I merely hold that these are not discernible at the most general level of analysis, and that the individual's acts or ideas do not *here and now* exert any influence on social, political or economic mechanisms."[52]

It can be argued that the autonomist assumption, that is, that society and culture are things in themselves and must be studied as such, is the most important single premise in all of sociology.[53] In this regard Ellul is distinctive only in the particular source he identifies at the heart of the condition of autonomy, *la technique*. He is also distinctive in his desire to pit the collective process against the pretensions of the individual. "Human beings are, indeed, always necessary," he agrees. "But literally anyone can do the job, provided he is trained to it."[54]

Ellul's treatment of technical self-augmentation is vulnerable to criticism on several points. In general, many of his conceptual distinctions are incorrect, and he seriously misinterprets what actually happens in scientific and technical development. His insistence that science and technique are identical simply does not wash. While it is true that the two spheres now have much in common, it is equally clear that in both conception and actual operation, they are still to a large degree separate kinds of enterprise. It seems wise, therefore, to maintain the basic distinction that science is a particular way of knowing or body of knowledge; technology is a particular kind of practice.

It is evident also that Ellul gives a misleading version of the way in which discoveries and inventions take place. His treatment suggests that the growth of scientific and technical knowledge is an entirely linear, rational, and additive accumulation. In this regard he is merely restating the image that scientists and technicians long held of their own work.

But if one reads the recent writings of historians and philosophers of science on this matter, one receives an entirely different picture. In the revised view, the advancement of knowledge in many fields has been far from a strictly logical, linear, or cumulative process. What one finds instead are paths marked by significant periods of turmoil, conflict, and radical shifts of orientation. Today the prevailing interpretations of the history of science stress the idiosyncrasies encountered in the progress of knowledge.[55] Studies of technological invention and innovation now pursue much the same direction.[56] At the same time, however, Ellul's conclusion about exponential development is not at all unusual. "What is true of science is true of technology," Donald Schon concludes. "Both have grown exponentially and the law governing their rate of change is the same as it has been for the last hundred or two hundred years."[57]

The self-augmentation conception can also be criticized for glossing over some important aspects of technical development itself. Ellul fails to notice any difference between invention and technical implementation and apparently believes that for all intents and purposes these activities are identical. Most of us, however, would want to point out that there is, after all, a distinction between what happens in the laboratory and what happens when discoveries are put to work in the world at large.

These and other objections of the same sort can be aimed at Ellul's formulation, some of them to good effect. Ellul does have several important facts wrong. His interpretation of several key issues is faulty. In his defense, it should be noted that most of the criticisms popularly raised against his work aim at relatively small points and do not come to grips with his central concern: to see the development of technology in its totality. We can be extremely proficient at questioning his assumptions and conceptual distinctions, but we usually avoid tackling his conclusions or their meaning. Often these conclusions and their general logic are formidable. In this sense much of what passes for criticism of the substance of Ellul's theory is more properly considered small potatoes quibbling about his methodology. All of Ellul's writing is informed by a desire to avoid the sterility of most modern thought, a sterility that comes from (among other sources) an unwillingness to risk an answer to the question, What *is* it that I see before me? While Ellul is obviously aware of the myriad of complex and delicate nuances bearing on his subject matter, he has elected to ignore many of them in order to proceed with his theory. Choices of this sort always involve a risk. But in taking that risk Ellul achieves something that is increasingly rare in modern social theory—the willingness to offer a complete, uncompromising statement.

What, then, is the significance of Ellul's position on self-augmentation? What questions of interest does it raise about technological change?

The aim of Ellul's discussion is to identify the origins of technological novelty. He is interested in the general circumstances which influence the introduction of new discoveries and devices into the world in an ever-expanding, ever-accelerating flood. What most people take for granted, Ellul finds an extraordinary state of affairs. Modern societies and their members must continually respond to scientific and technological innovations for which they have little or no preparation. Yet the tide continues, essentially unplanned and unabated, with neither limit nor preconceived direction. In recent years a number of scholars have joined Ellul in taking wonder at this situation, but most of them have seen the issue as one of personal and social adaptation. Few have wanted to do what Ellul insists must be done: to notice the elements of dynamism, necessity, and ineluctability built into the *origins* of the process.

As we have seen, the dynamism that Ellul uncovers at this level is based on a series of contingencies. It is based on an if . . . then kind of logic, with an attempt to demonstrate that all of the "ifs" actually hold. The self-augmentation of technique results from the convergence of three conditions: (1) the universal willingness of people to seek and employ technological innovations, (2) the existence of organized social systems in all technical fields, and (3) the existence of technical forms upon which new combinations and modifications are based. If all of these elements are present, the rapid expansion of technique is assured.

Perhaps the most interesting implication of Ellul's case is that each specific invention or discovery in a sense "has its time," a time when its introduction into the world becomes virtually (though not absolutely) inevitable. Each new advance is latent in the existing technical ensemble. Given the social system of inquirers and innovators, it will somehow find its way into the world. Someone will find that "it works this way" or "this is how it fits together."

George Kubler advances a similar point, in his provocative interpretation of the history of art. Even the most original pieces, he argues, are more or less similar to artifacts of the same sort (painting, sculpture, pottery, or some other) that came before. The life of an artist and the character of his or her works must, therefore, be seen as a matter of their position in a continuing sequence of lives and works. "When a specific temperament interlocks with a favorable position," Kubler explains, "the fortunate individual can extract from the situation a wealth of unimagined consequences. This achievement may be denied to other persons, as well as to the same person at a different time. Thus every birth can be imagined as set into play on two wheels of fortune, one governing the allotment of its temperament, and the other ruling its entrance into a sequence."[58]

As evidence for his case, Ellul cites a peculiar situation in modern science and technology: the frequent occurrence of simultaneous discoveries and inventions. Anthropologist A. L. Kroeber was also fascinated with this state of affairs and drew conclusions similar to Ellul's. His research revealed that many important advances—among them the telegraph, telephone, photography, the periodic law of elements—had come from the work of two or more individuals working independently at approximately the same time. In Kroeber's eyes this tended "to instill a conviction that inventions may be inevitable, within certain limits; that given a certain constellation and development of a culture, certain inventions must be made."[59] Kroeber did not believe that this discovery need dampen the spirit of scientists and technicians. They would, after all, still receive credit for their work from their peers. But in a larger sense, the cultural sense, "it is only a question of who will work the idea out feasibly. Will it be Bell or Gray in 1876, or someone else in 1877 or 1878 or perhaps as early as 1875? To the individual inventor the 'Who' is all-important, because it means who is to get the prize. To his society, and to the world at large, the 'Who?' is really a matter of indifference—except for sentimental partisanship—because the invention was going to be made anyway about when and where it was made."[60] Kroeber, furthermore, was not the least bit shy in labeling this situation a variety of determinism: "In a familiar metaphor, we say that the discovery is now in the air, or that the time is ripe for it. More precisely, inventions are culturally determined. Such a statement must not be given a mystical connotation. It does not mean, for instance, that it was predetermined from the beginning of time that type printing would be discovered in Germany about 1450, or the telephone in the United States in 1876. Determinism in this connection means only that there is a definable relation between a specific condition of a given culture and the making of a particular invention."[61]

We are justified in asking Ellul and Kroeber whether there is any substance to this determinism. Is it not possible to pose the same objection that we placed at the door of the technoevolutionists: that they had actually assumed what they later discovered—the insignificance of the human being in the multiplication of technics?

To be more specific, we must ask whether the kind of involuntarism found here is merely a product of Ellul's chosen theoretical perspective. In his argument, the rate of technical progress becomes a "social fact" in much the same way that the suicide rate was a social fact for Durkheim. Individual inventions or individual acts are of little interest. The status of particular deeds and of their voluntary character is discounted. In laying the foundations of his sociological method, Durkheim put forth the following rule: "*The voluntary character of a practice or an institution should never be assumed before hand.*"[62] After examining cases in which this rule and others would be useful in the study of society, he went on to state the principle more forcefully: "*The determin-*

ing cause of a social fact should be sought among the social facts preceding it and not among the states of individual consciousness."[63]

But for Ellul, as for Durkheim, the withering of freely made decisions and actions within the social fact is never merely a methodological convenience. A social fact is truly a *fact*. It has reality of its own and imposes a kind of necessity upon people in society. Ellul, following Durkheim, is not merely suggesting where or how to look at something. He is trying to get at the truth of the matter. For the sake of argument we may grant Ellul's proviso that in most cases the individuals involved will not perceive the context of necessity in which they exist. But if the self-augmentation of technique is a *fact*, then we would expect that it would have been noticed, at least on occasion, by those caught up in its forward sweep. Are there any such cases?

Ellul's account is one way of thinking about a condition that has come to plague many scientists and technicians in the mid-twentieth century. Sensing that the fruits of their labor may well have pernicious implications for the future of the human race, these people have also realized that there is really no choice but to continue. Whether positive or negative in its implications, the development at hand *will* enter the world.

The poignancy of this state of affairs is etched on the pages of Werner Heisenberg's *Physics and Beyond*. In a chapter on the responsibility of the scientist, Heisenberg tells of the day in 1945 when he and his fellow German scientists, then held in captivity by the Allies, learned that an atom bomb had been dropped on Hiroshima. "I had reluctantly to accept the fact," he recalls, "that the progress of atomic physics in which I had participated for twenty-five long years had now led to the death of more than a hundred thousand people."[64] In the passage that follows, Heisenberg reconstructs from memory the conversations he had with the other scientists held prisoner—Otto Hahn, Max von Laue, Walther Gerlach, Carl Friedrich von Weizsäcker, and Karl Wirtz. "Worst of all," Heisenberg remembers, "was Otto Hahn. Uranium fission, his most important scientific discovery, had been the crucial step on the road to atomic power." Hahn immediately retreated to his quarters, "visibly shaken and deeply disturbed, and all of us were afraid that he might do himself some injury."[65]

In Hahn's absence, Carl Friedrich von Weizsäcker spoke of the quandary in which the scientists found themselves. "It is easy to see why Hahn should be dejected. His greatest scientific discovery now bears the taint of unimaginable horror. But should he really be feeling guilty? Any more guilty than any of us others who have worked in atomic physics? Don't all of us bear part of the responsibility, a share of his guilt?"[66]

"I don't think so," Heisenberg replied. "The word 'guilt' does not really apply, even though all of us were links in the causal chain that led to this great tragedy. Otto Hahn and all of us have merely played our part in the development of modern science. This development is a vital process, on which mankind, or at least European man, embarked centuries ago—or, if you prefer to put it less strongly, which he accepted."[67]

The physicists went on to affirm their belief in the essential goodness of scientific knowledge and to ponder whether a world government might bring the matter of right or wrong ends under control. In Carl Friedrich's words, "Our task, now as in the past, is to guide this development toward the right ends, to extend the benefits of knowledge to all mankind, not to prevent the development itself. Hence the correct question is: What can the individual scientist do to help in this task; what are the precise obligations of the scientific research worker?"[68]

Heisenberg's answer to this question, based on his conception of how science grows, is an interesting one. He denies that the individual scientist *can* have any special responsibility for his discoveries.

"If we look upon the development of science as an historical process on a world scale," I replied, "your question reminds me of the old problem of the role of the individual in history. It seems certain that in either field the individual is replaceable. If Einstein had not discovered relativity theory, it would have been discovered sooner or later by someone else, perhaps by Poincaré or Lorentz. If Hahn had not discovered uranium fission, perhaps Fermi or Joliot would have hit upon it a few years later. . . . For that very reason, the individual who makes a crucial discovery cannot be said to bear greater responsibility for its consequences than all the other individuals who might have made it. The pioneer has simply been placed in the right spot by history, and has done no more than perform the task he has been set."[69]

The issue here is not merely the one most frequently noted, that scientific discoveries and technological advances have a capacity for both benefit and harm. It is, instead, that there is evidently an element of virtual necessity or inevitability at work here. "The worst thing about it all," Heisenberg tells his companions, "is precisely the realization that it was all so unavoidable." One might suppose that at the level of invention or discovery the choice is always open to suppress the development. But as Carl Friedrich correctly points out, this is not the scientist's job. "Our task is . . . not to prevent the development itself." And there is, indeed, little evidence in the history of science and technology of anyone suppressing his work because he believed it might have adverse consequences for humanity.[70] Heisenberg's contribution is to show that even if such feelings did exist, it would make no difference. Someone else is bound to introduce the knowledge or device into the world. The scientist's only choice is to enter the moral and political discussion that follows upon the development, and here his voice is just one among the multitude. "He may possibly be able to exert just a little extra influence on the subsequent progress of his discovery, but that is all. In fact, Hahn invariably made a point of speaking out in favor of the exclusive use of uranium to peaceful purposes . . . of course, he had no influence on developments in America."[71]

Heisenberg's thinking here is neither unique nor an aberration. We can find similar sentiments in much of the contemporary literature on the social responsibility of the scientist. In Norbert Wiener's *Cybernetics*, the author ponders the implications of his work in communications theory and sinks into a melancholy mood.

Those of us who have contributed to the new science of cybernetics thus stand in a moral position which is, to say the least, not very comfortable. We have contributed to the initiation of a new science which, as I have said, embraces technical developments for good and evil. We can only hand it over into the world that exists about us, and this is the world of Belsen and Hiroshima. We do not even have the choice of suppressing these new technical developments. They belong to the age, and the most any of us can do by suppression is to put the development of the subject into the hands of the most irresponsible and most venal of our engineers . . . there are those who hope that the good of a better understanding of man and society which is offered by this new field of work may anticipate and outweigh the incidental contribution we are making to the concentration of power (which is always concentrated, by its very conditions of existence, in the hands of the most unscrupulous). I write in 1947, and I am compelled to say that it is a very slight hope.[72]

There are times when a choice cannot be made and when the drift of events cannot be halted. This is the irony that Ellul describes: Men are free in one respect but totally boxed in in another. At the points of origin, technology moves steadily onward as if by cause and effect. This does not deny human creativity, intelligence, idiosyncracy, chance, or the willful desire to head in one direction rather than another. All of these are absorbed into the process and become moments in the progression.

The statements of Heisenberg and Wiener mirror the myth of Pandora's box or, perhaps more fittingly, Greek tragedy itself. What we witness is the predicament of those who look on helplessly as their deeds contribute to what may well be their own downfall and, possibly, the downfall of the entire community. An atom-space-computer age Moira is at work here—a fate that employs the free action of men to bring about ends that carry an aroma of necessity.

One need not look to the great luminaries of science and technics to hear this story told. Inertia in technological evolution is now experienced as a problem at a much more mundane level, the level at which most people live and work. Announced by such terms as "future shock," "temporary society," and "occupational obsolescence," the fear continues to grow that the pace of advancing technology makes it difficult for a person to keep up with the techniques required for his work or to adapt to the technical changes that again and again revolutionize basic conditions of social existence. These anxieties are actually much closer to what Ellul tries to express—the never-ending, ever-accelerating expansion of technics in every field. While the technoevolutionist view of obsolete homo sapiens seems ill founded, the possibility of human obsolescence in everyday work is very real indeed.

To search for schools of thought that offer an alternate course to this onward momentum is futile. There are none. Any suggestion that there be a slowdown, limitation, or moratorium on scientific inquiry, research, and development is unthinkable at present. By general consensus, one may not tamper with the source. Faced with the possible dangers to human life involved in their research on DNA, an international group of molecular biologists called a moratorium on a part of their work in 1974 and 1975. It was the first such moratorium in memory, and, predictably, it was short-lived. Existing safeguards, the scientists finally decided, were adequate to handle the dangers. But at meetings held on the subject, the chilling point was frequently raised that the technical capability required to do the potentially perilous research might soon be within the reach of high school chemistry students, not to mention unscrupulous investigators uninterested in ethics or safeguards. Nobel Prizes are not won by suppressing research techniques or fruitful lines of inquiry. The argument, "If we don't do it, somebody else will," cuts in more than one direction.

What Ellul labels "self-augmentation," then, far from being a mysterious process, is a reflection of the ongoing work of thousands of individuals in our scientific and technical communities supported by some of our most cherished beliefs. None of this implies that outcomes are inevitable in any absolute sense. As Kubler observes, "The presence of the conditions for an event does not guarantee the occurrence of that event in a domain where a man can contemplate an action without committing it."[73] But for whatever reason, the sphere of technological development tends not to be one in which human motives are scrupulously clarified or where prudence reigns. In hearings before the Personnel Security Board in 1954, J. Robert Oppenheimer was asked whether it was true that his qualms about whether the United States ought to develop the hydrogen bomb increased as the feasibility of the device became clearer. He answered: "I think it is the opposite of true. Let us not say about use. But my feeling about development became quite different when the practicabilities became clear. When I saw how to do it, it was clear to me that one had to at least make the thing. Then the only problem was what would one do about them when one had them. The program in 1949 was a tortured thing that you could well argue did not make a great deal of technical sense. It was therefore possible to argue also that you did not want it even if you could have it. The program in 1951 was technically so sweet that you could not argue about that."[74]

Evidently technological accomplishment has become a temptation that no person can reasonably be expected to resist. The fact that something is technically sweet is enough to warrant placing the world in jeopardy.

Technological Determinism

The root of our word *invention* means "to come upon," while that of *innovation* points to something that "renews or makes new." Today the relationship between things freshly come upon and things made new is taken for granted; the link between discovery in science, invention in technics, and innovation in society seems automatic. There have been times, however, when this connection did not hold. In ancient Greece, in Hellenistic civilization, and in Islam of the Middle Ages we find cultures in which highly sophisticated scientific and technical knowledge existed without a profound effect on social practice. The Alexandrian inventors of the second and third centuries B.C. worked with devices which were to play an important role in modern European industry. Hero of Alexandria, for example, developed primitive versions of the basic elements of the railroad—the steam engine and a wheeled cart that moved along on wooden rails. But in the setting of the period, such inventions had little utility. Science remained an isolated realm of contemplation. Experimental technics became an activity that aimed at producing little more than ingenious curiosities. The only use found for Hero's marvelous railway was the hand-driven deus ex machina, a deity in Greek and Roman drama wheeled in to influence the action taking place on stage. His steam engine prototype, "The Sphere of Aeolus," was never more than an interesting toy. Pondering the eventual paralysis of Greek science and technics, Benjamin Farrington concludes, "The ancients rigorously organized the logical aspects of science, lifted them out of the body of technical activity in which they had grown or in which they should have found their application, and set them apart from the world of practice and above it."[75]

In modern society men and institutions readily embrace inventions and discoveries and take steps to see that they quickly become innovations in the broader sphere of practical activity. No longer do we find that an invention waits for decades or centuries before entering an active role in civilization. The phenomenon of rapid, seemingly ubiquitous innovation provides an important set of questions for the idea of autonomous technology, for changes in technics bring about changes in other things as well. Whereas the immediate application of a particular technology is usually conscious and deliberate, other consequences of its presence in the world often are not. It is this gap between original intentions on the one hand and ultimate effects on the other, between the truly chosen and the never chosen, that has perplexed many schools of thought in our time.

To speak of technological change in terms of its broader effects raises a question that can easily become a swamp of intellectual muddles: the doctrine of technological determinism. Often couched in the noncommittal language of "the impact of technological innovation," the idea plays a prominent role in a great deal of contemporary writing on technology and society. One need only look at the literature on "technology assessment," "social indicators," "alternative futures" or "the year 2000" to see uncriticized quasideterministic assumptions at the center of speculation and research.[76] The works of Daniel Bell, Herman Kahn, Olaf Helmer, Raymond Bauer, and other notables are clearly informed by the belief that alterations in technology have been and will probably continue to be a primary cause of change in our institutions, practices, and ideas. Very often the issue is dodged by focusing on a single development in technics and observing its "impact." A scholar demonstrates how satellite communications will have a tremendous influence on our social or political life or how "fundamental changes in the socio-economic system . . . are being brought about through drives exerted on the whole social fabric by the applications of cybernetics in the form of computerized systems."[77] But the implications of these views or the general theory which could make them meaningful in a world in which satellites and computers are just one segment of an expanding technical complex are not considered.

In a fundamental sense, of course, determining things is what technology is all about. If it were not determining, it would be of no use and certainly of little interest. The concept "determine" in its mundane meaning suggests giving direction to, deciding the course of, establishing definitely, fixing the form or configuration of something.[78] The first function of any technology—and the immediate condition of its utility—is to give a definite, artificial form to a set of materials or to a specific human activity. To put it differently, it provides structure for the primary medium to which it is applied. An apparatus is a structure of material parts so assembled as to produce determinate, predictable results when placed in operation. A technique is a structure of human behavior designed to accomplish a definite outcome. A technical organization is an assemblage of human beings and apparatus in structured relationships designed to produce certain specified results. A technical operation, to the extent that one engages in it, determines what one does. If the operation is successful, we may say that the technology determined the result. This does not mean that either the technology or its result are totally rigid or inflexible. What it does indicate is that technology succeeds through the conquest of disorder and the imposition of form.

Controversies on the matter, however, stem from a much broader and more dubious notion. Understood in its strongest sense, technological determinism stands or falls on two hypotheses: (1) that the technical base of a society is the fundamental condition affecting all patterns of social existence and (2) that changes in technology are the single most important source of change in society. In this form very few thinkers have been willing to adopt an unabashedly determinist position. One of the intrepid, anthropologist Leslie White, explains: "We may view a cultural system as a series of three horizontal strata: the technological layer on the bottom, the philosophical on the top, the sociological stratum in between. These positions express their respective roles in the culture process. The technological system is basic and primary. Social systems are functions of technologies; and philosophies express technological forces and reflect social systems. The technological factor is therefore *the* determinant of a cultural system as a whole. It determines the form of social systems, and technology and society together determine the content and orientation of philosophy."[79]

The standard objections to this view emphasize both a methodological and a moral point. On the one hand there is the problem central to any cultural or social science of isolating "causes." It is almost impossible to single out any one factor as the origin of the changes that must be explained. The idea that technology or anything else could be the primary determinant is impossible to prove. Patterns of technology are themselves largely influenced by the conditions of the societies in which they exist.[80] The kinds of tools and technical arrangements found in Japan bear the distinctive stamp of that nation's cultural traditions and will be different from technologies with the same approximate functions in Italy or the United States.[81] The character of a society and the changes that take place in it are the product of a vast set of possible causes—climate, geography, population, religious practices, the market, political structure, and so forth. Within the state of our present knowledge, it is not possible to demonstrate conclusively that technology or any other single factor is most important.

Along with this come objections which the voluntaristic position raises. The determinist doctrine offends our sense that fundamental technological conditions are chosen and that social forms related to technology are not merely the passive imprint of new varieties of apparatus or technique. "As our understanding of the history increases, it becomes clear that a new device merely opens a door; it does not compel one to enter."[82] Such assertions are generally offered as reassurance that society is, after all, free to choose and that it is a mistake to look for elements of determinism in the course of technological change.

But although fraught with difficulties, the idea of determinism is not one that ought to be rejected out of hand. The tendency to dismiss the entire issue after scoring a single moral or methodological point places a taboo on important questions that even a cursory glance at modern history suggests are among those most crucial to an understanding of our age. Looking at specific changes—the building of the railroads, electrification, Taylorism, mass communications—it is clear that there are many ways in which technics has shaped the specific forms of modern life. A vague multifactoralism and glib reassertion of the voluntarist position bestow a haphazard, soft-headed quality to much of the historical scholarship and social theory on this question. William Ogburn's cultural lag thesis, whereby social and cultural institutions are seen to drag behind technological development by a number of years, is still, unfortunately, just about the most profound general view that we have.[83]

The task, then, is to avoid throwing out the baby with the methodological bath water. To help us decide what is living and what is dead in the discussion about determinism there is no better place to begin than a reexamination of the social theorist who was first to give the matter systematic, rigorous attention. In many ways the writings of Karl Marx are still unsurpassed in their clarity and insight on this theme.

Marx's conception of technical and social change is an aspect of his general theory of historical materialism. The theory, an attack upon the philosophical and political idealism of his time, argues that people do not establish their conditions of life or their identities through spontaneous, preconceived ideas. "What individuals are . . . ," he asserts, "depends on the material conditions of their production."[84] As noted earlier, Marx holds that human beings are necessarily involved in sensuous, productive activity in a world of material things. Since life and production are virtual equivalents, the mode of production present in a society takes on a supreme significance. "This mode of production," Marx explains, "must not be considered simply as being the reproduction of the physical existence of the individuals. Rather it is a definite form of activity of these indiviuals, a definite form of expressing their life, a definite *mode of life* on their part. As individuals express their life, so they are. What they are, therefore, coincides with their production, both with *what* they produce and with *how* they produce. The nature of individuals thus depends on the material conditions determining their production."[85]

It is upon this view of man, life, and activity that the deterministic elements in Marx's theory are based. A famous summary statement in the preface to *The Critique of Political Economy* indicates how broadly he understands the workings of this historical principle: "The mode of production of material life determines the general character of the social, political and spiritual processes of life."[86]

In other writings Marx goes on to separate the notion of mode or conditions of production into two distinct categories: forces of production and relations of production. Although there is some variation in the manner in which Marx uses these terms, for our purposes "forces of production" can be understood to comprise all of physical technology. Marx usually employs the term to mean the sum total of tools, instruments, and machines available in a society, plus all sources of energy that move these implements: steam, water, coal, animal, and human power. By selecting the term *forces of production (Productivskräfte)* he brings together under a single concept the instruments, energy, and labor involved in the active effort of individuals to change material reality to suit their needs.

Again and again in his writing Marx states that the forces of production play a determining role in human history. There are indeed other factors present as well, but by far the most important influences upon the shape and content of social existence are those exerted by the forces of production.[87] In *The German Ideology* we read "that the multitude of productive forces accessible to men determines the nature of society, hence that the 'history of humanity' must always be studied and treated in relation to the history of industry and exchange."[88] *The Poverty of Philosophy* states: "In acquiring new productive forces men change their mode of production; and in changing their mode of production, in changing the way of earning their living, they change all their social relations. The handmill gives you society with the feudal lord; the steam-mill, society with the industrial capitalist."[89]

Insofar as there is significant change in society, it is usually because there has been a change in the forces of production. In the language of social science, Marx has isolated *the* primary independent variable active in all of history. His illustrations of how it affected the course of human development are drawn from the whole of the historical record. Pointing to the changes which brought the "feudal system" to Europe, for example, he concludes, "To what an extent this form was determined by the productive forces is shown by the abortive attempts to realize other forms derived from reminiscences of ancient Rome (Charlemagne, etc.)."[90] All of men's conscious designs that run athwart the productive forces are inevitably frustrated.

Further elements of determinism in Marx's theory can be seen in his view that within each historical period the social relationships found within the productive process are a function of the forces of production existing at the time. He argues that "a change in men's productive forces necessarily begins a change in their relations of production."[91] New relations are formed and older ones are destroyed. In describing this situation Marx is emphatic about the direct connection between forces of production and what he terms "the division of labor." As forces of production develop, as they become more numerous, more sophisticated, more powerful, they necessitate a further development in the division of labor. There is an increase in the number of technical and financial operations, an increase in the complexity of the social organization of work, and a specialization of the skills required in all labor. "How far the productive forces of a nation are developed is shown most manifestly by the degree to which the division of labor has been carried. Each new productive force, in so far as it is not merely a quantitative extension of productive forces already known, (for instance the bringing into cultivation of fresh land), brings about a further development of the division of labour."[92] The introduction of machinery in modern industry had, for example, led to a further development of the social division and organization of labor. While in most cases Marx seems to be saying that there is a one-way influence between the forces and relations of production, he also notes that there is a circular process of development occurring here. "As the concentration of instruments develops, the division develops also, and *vice versa.* This is why every big mechanical invention is followed by a greater division of labour, and each increase in the division of labour gives rise in turn to new mechanical inventions."[93]

It would be a mistake, however, to interpret Marx's treatment of such "relations" and "divisions" in any narrow manner. His analysis can indeed be used to think about specific industrial operations on a limited scale. It can help us understand what would be socially necessary to operate a particular factory given a certain kind of machinery. But his categories are also meant to be understood in a very general sense. Georg Lukács is correct in pointing out that Marx's intention is to describe a "totality."[94] In the end, the relations of production consist of all organized relationships in society. The division of labor includes every existing division in socially necessary, productive labor throughout the whole social system. The forces of production determine the full scope of these divisions and relationships. In Marx's words, "With the acquisition of new productive faculties, men change their mode of production and with the mode of production all the economic relations which are merely the necessary relations of this particular mode of production."[95]

Seen in this way, purely technical relationships and economic relationships amount to the same thing. "Division of labor and private property are, moreover, identical expressions: in one the same thing is affirmed with reference to activity as is affirmed in the other with reference to the product of activity."[96] From the relations of production and division of labor spring property, ownership, social class, and, ultimately, the consciousness of persons in each class. In a rough approximation one can say that the available productive forces "cause" the shapes of sociotechnical organization, ownership, and class structure one finds in society. But a more accurate statement would be that these social configurations are those both appropriate to and necessary for the productive forces at hand.

In summary, Marx begins with the notion that the mode of production forms a whole and that it is a *mode of life*. That which determines production determines the life of a society in its full spectrum. This comes about not as the result of remote causes but rather as a consequence of willful, active human individuals responding to what is both possible and necessary in their time. It is, however, above all else the available technology which enforces limits upon the possible and the necessary. History reveals the existence of social classes vastly unequal in wealth and power, formed in a struggle defined by the conditions of material production.

There is a point at which Marx simply assumes the determining influence of "the technical composition of capital" and begins a full-scale study of the labor theory of value, the accumulation of capital, and so on. For our purposes here, we will not follow him in that direction. The standard categories of economics, whether in Marxian or liberal form, tend to assume too much and reduce technological issues to a few abstract variables that render invisible a great expanse of interesting territory. Despite its undeniable merits, the approach taken by the economist tends to obscure many pressing questions, including the one before us now: technological determinism and freedom.

In its most extreme aspect, of course, Marx's work aims at devising a rigorous science. Historical materialism in the scientific sense set out to discover deterministic laws of social change and to make accurate predictions based on such laws. It was this feature of Marxism that Friedrich Engels tried to carry beyond what Marx had done, a feature that attracted many nineteenth- and twentieth-century followers. This attempt at science, best seen in Engel's *Dialectics of Nature* and *Anti-Dühring,* is proudly deterministic as it views events in nature and history and as it regards the possibility of human freedom. Bound by laws of historical materialism, individuals, classes, and societies have little autonomy to guide their own character or development.[97]

By the standards normally applied to scientific work, historical materialism fares very poorly. Its methodology in accounting for the allegedly dialectical changes taking place in society is at best highly suspect. Its futility in formulating and testing objective, predictive laws is notorious. Increasingly, Marxism finds its adherents among those interested in its richness as a philosophy or mode of social analysis rather than in its more ambitious scientific claims. The failure of the deterministic science, however, cannot be taken as sufficient to reinstate voluntarism. Marx's rich insights into the forces that shape human action do not depend on the ability to produce a science of society as reliable as physics. Taking his work as a point of reference, one should now be able to summarize briefly a set of claims that, though less grand than those sometimes made, constitutes a reasonable notion of technological determinism and the problems it presents for human freedom.

To emphasize what should already be clear, Marx's view does not consider individuals helpless automatons obeying clockwork patterns or passively receiving the imprint of history's stamping press. His sense of the human being as an active, productive agent carries through to his conception of change. "It is not 'history' which uses men as means of achieving—as if it were an individual person—its own ends. History is nothing but the activity of men in pursuit of their ends."[98] But this does not mean that men are totally free to select what they will do or the conditions under which they will live. To the contrary, it is always the case that individuals are severely limited in what they can choose. Marx again and again derides the view that sees history as an open and infinite set of opportunities waiting to be seized upon by men of imagination, pluck, and luck. In particular, men are not at liberty with regard to the forces of production with which they must live. These are, for the most part, received as the accumulated product of discoveries, inventions, and everyday work from previous generations.

> It is superfluous to add that men are not free to choose their *productive forces*—which are the basis of all their history—for every productive force is an acquired force, the product of former activity. The productive forces are therefore the result of practical human energy; but this energy is itself conditioned by the circumstances in which men find themselves, by the productive forces already acquired, by the social form which exists before they do, which they do not create, which is the product of the preceding generation. Because of this simple fact that every succeeding generation finds itself in possession of the productive forces acquired by the previous generation, which serve it as the raw material for new production, a coherence arises in human history.[99]

To speak of "determinism" in circumstances of this kind is perhaps to put the case too strongly. Marx calls our attention to the fact that each generation is strongly *conditioned* or informed by a technological inheritance that it in no sense "chose." While it is always possible that a particular generation might wish to review this inheritance, scrutinize the patterns that technics gives to life, and make new choices on the basis of this critique, such a procedure is not in fact something that occurs to anyone to do. In the main, the sociotechnical context into which we are born must simply be accepted as given. "Means of production," Marx concludes, "do not depend on free will."[100]

Another important theme of determinism in Marx's theory focuses upon the way in which technologies give structure to human needs. In Marx's view, needs are not present in any simple, finished form in man's biological composition. Instead they are relative to and change with the condition of society at a given time and at a particular stage in the forces of production. The *Grundrisse* observes that "needs were slight in the beginning, and only developed with the productive forces."[101] Marx exalts capital because it propels civilization beyond "self-sufficient satisfaction of existing needs confined within well-defined bounds, and the reproduction of the traditional way of life. It is destructive of all this, and permanently revolutionary, tearing down all obstacles that impede the development of productive forces, the expansion of needs, the diversity of production and the exploitation and exchange of natural and intellectual forces."[102] As technology advances, human needs actually "expand" and change in quality. Potentially, he believes, this tendency toward growth and development in human motives is a limitless process.[103]

The concept of *need* contains at least two meanings. It signifies both a notion of *necessity*—things wanted because they comprise conditions for survival or basic human existence—and a notion of *desire*—things not strictly necessary but wanted for the satisfaction they bring. If one begins with the supposition that the human being has as part of its basic character a general unformed set of urges, then the technological determinism of need becomes a powerful hypothesis. Humans require such things as food, shelter, sex, and the opportunity to communicate with their fellows. But once they adopt a particular technical form in pursuit of these ends, the context of both necessity and desire becomes highly specific. At a certain stage in the development of technics, the need for physical mobility actually becomes the need to have access to automobiles, airlines, or effective equivalents. Such needs are as basic for that stage of technical capacity as the need for oxen or a good pair of sandals might have been for an earlier one. The development of productive forces not only generates variations on older needs but in a true sense creates whole new ones. Thus, the feeling that soap is one of life's necessities appeared only with the coming of industrial techniques of soap manufacture. With the spread of this innovation came an unprecedented desire for a well-scrubbed cleanliness that is now second nature to most of us. There have been times and cultures, however, in which our need to do away with dirt, "germs," and odors would have seemed totally puzzling.

Marx's conception of technological innovation and its relationship to change in social forms and human motives presents a direct challenge to voluntarist notions of choice. Nowhere in his theory are free, deliberate decisions excluded as a logical possibility. There is no metaphysical argument that our universe is such that freedom cannot exist. What

Marx brings us to question, however, is the matter of when and where something properly called a "choice" comes up at all. His conclusion is that the arena here is actually quite small. History moves by a much less self-conscious, deliberate process with regard to forces of production, a process that adds up to a kind of continuing adaptation. Opportunities for willful, conscious choice about social patterns linked to the forces of production never arise. That is not the way change occurs. "History shows," he argues, "how the division of labour peculiar to manufacture, strictly so called, acquires the best adapted form at first by experience, as it were behind the backs of actors, and then like the guild handicrafts, strives to hold fast that form when once found, and here and there succeeds in keeping it for centuries. Any alteration in this form, except in trivial matters, is solely owing to a revolution in the instruments of labour."[104]

Marx's contention that changes take shape "behind the backs of actors" may seem contradicted by the fact that certain persons obviously do make decisions and choices in the process. Inventors or those in today's business of research and development do have the opportunity to choose among a variety of possible paths that technics might take. A selection can be made between a new automobile assembly line run by automation or by a predominance of manual labor. Similarly, entrepreneurs have the chance to make certain choices at other junctures. Thus, for example, Aristotle Onassis did decide to build a fleet of supertankers. Those who designed and built the ships made choices about their physical characteristics and the social organization of work on board.

But before deciding the issue, Marx would have us consider the total range of effects in human life over a long period of time occasioned by the assembly line, supertanker, or any other significant development in forces of production. One aspect of the truth, surely, is that a relative few are able to make choices that bind the many. But even beyond this, as we look to the entire process of change, is a vast series of partial adaptations adding up to results that no one "chose" or "controlled." For example, when a badly constructed supertanker breaks up on the shoals, spreading oil on beaches, we must understand that the event has something to do with decades of technical and social change that created the circumstances for the calamity. But does it make sense to say, as the voluntarist argument suggests, that we "chose" the design of the ships, the form of Onassis's corporation, the social and political conditions under which the boats sail, or the eventual crack-up? When we think back on it, we do not remember having been consulted.

Much of our ordinary contact with things technological, I would argue, is of exactly this kind. Each individual lives with procedures, rules, processes, institutions, and material devices that are not of his making but powerfully shape what he does. It is scarcely even imaginable what it would mean for each of us to make decisions about the vast array of sociotechnical circumstances that enter our experience.

A powerful illustration of the issues of determinism and choice in technological development is provided by Pertti J. Pelto's study of the changes brought to the Skolt Lapps of Finland by the introduction of the snowmobile into their society in the 1960s. In many ways the story reflects in miniature the whole course of the industrial revolution. The Lapps of the Sevettijärvi region deliberately *chose* to use snowmobiles as a replacement for dogsleds and skis in the basis of their economy—reindeer herding. But they neither chose nor intended the effects this change would have in totally reshaping the ecological and social relationships upon which their traditional culture depended.[105]

First introduced to the Sevettijärvi region of Finland by a schoolteacher who used it on fishing expeditions, the snowmobile was soon adopted by the Skolts as an important innovation in their work. The speed of the machines made it possible to do the annual reindeer roundup in a single operation. As a consequence, the age-old winter herding practices that kept the herdsmen and animals in close contact became unnecessary and were abandoned. Under this arrangement previously "tame" reindeer returned to the wild and could be corralled only by a blitz of organized snowmobiles. While there were obvious economic benefits that resulted from the use of the new machines, a number of lamentable side effects were soon noticed. Possibly a result of physiological strain placed on pregnant female reindeer by the stampede running of mechanized roundups, the fertility and population of the herds fell off sharply. The absolute decline in number of animals was compounded by a marked decrease in the number of Skolt families able to participate in snowmobile-based herding. Faced with the cash squeeze brought by the purchase and upkeep of the machines, many families were forced to sell their stock and eventually dropped out of the business altogether.

As Pelto observes, there emerged groups of clearly recognizable "winners" and "losers" in this process. Some were better able than others to adopt the new techniques, master the skills necessary to succeed in a money economy, and build a strong position in a new set of social relations. Those steeped in the old ways of knowledge and practice suffered the embarassment of being "unemployed," a role previously unheard of in the traditional culture, or of swallowing the herdsman's pride and becoming wage laborers. The development of a more complex system of socioeconomic differentiation was accompanied by the rise of "needs" associated with a more modern style of life—washing machines, household gas, telephones, and chain saws. How successfully a Skolt adapted to the snowmobile determined, by and large, his access to these goods. Hence, what had previously been a highly egalitarian society became inegalitarian and hierarchical almost overnight.

From one point of view the Skolts knew exactly what they were doing. They adopted the Bombardier "Ski-Doo" to make herding faster and more efficient. From another point of view, however, they never knew what hit them. The changes they saw taking place in long-established patterns of life just "happened" as the community made a place for this new instrument of production. Pelto's account acknowledges both deterministic and voluntaristic aspects of the developments he observed. "The evidence is strong," he writes, "that the introduction of a new technological device in a socioeconomic system has produced very extensive direct and indirect modifications of work patterns, household maintenance systems, and other aspects of adaptive behavior. At the same time we are confronted with the inescapable importance of *individual differences* (in both physical and psychological characteristics) as factors affecting the adaptive strategies that are played out in the ongoing social action."[106] Pelto finds that neither the model of "techno-economic determinism" nor that of "cultural causation" is adequate to explain changes of the kind the Skolts experienced. But his judiciousness here does not lead him to take a step for which others who study technological change often yearn—the secure retreat to a purely voluntarist conception. Strongly suggested in his study is the crucial difference between *choices*, properly considered, and *adaptive responses* to the conditions brought by a new order.

In short, without falling prey to the tortured dilemmas of the idea of causation or the mistaken quest for a materialist science of history, one can reconstruct a reasonable notion of what the determinist view wants to say. Glorious exhortations in contemporary writing—"Technological man will be his own master,"[107] "Technological man will be man in control of his own development within the context of a meaningful philosophy of the role of technology in human evolution."[108]—are unmasked as mere phantoms of the imagination. As Marx counsels, human beings do make their world, but they are also made by it.

Technological Drift: Uncertainty and Unintention

The picture of technological change that begins to emerge from our discussion is not that of a law-bounded process grinding to an inevitable conclusion. It is rather that of a variety of currents of innovation moving in a number of directions toward highly uncertain destinations. In what has become a standard reformulation of the problem, Robert Heilbroner concludes, "Technological determinism is . . . peculiarly a problem of a certain historic epoch—in which forces of technical change have been unleashed, but when the agencies for the control or guidance of technology are still rudimentary."[109] Perhaps the appropriate label for this state of affairs is not determinism at all but, instead, technological drift.

In a widely diverse collection of contemporary writings, the story is told in the following way. A multiplicity of technologies, developed and applied under a very narrow range of considerations, act and interact in countless ways beyond the anticipations of any person or institution. Except in cases of extreme danger or disaster, there are almost no existing means for controlling or regulating the products of this chain of events. People still retain their logical postition as users and controllers of technology. But in the broader context which transcends both "use" and "control," this logic is of little consolation. As the speed and extent of technological innovation increase, societies face the distinct possibility of going adrift in a vast sea of "unintended consequences."

This litany of determinism via indeterminism, necessity through aimless drift, has some important implications for contemporary political thought. If one compares the notions of pluralist politics to our ideas of technology-linked change, one finds an interesting disjunction. In the American model of pluralism founded by Madison and perfected in our time by David Truman, Robert Dahl, and others, the life of the polity rests in the many interest groups that serve as guardians of the welfare of their constituencies. The theory is based on the idea that individuals perceive their own interests, organize around concerns that affect their well-being, voice their desires through these organizations, and battle other similarly organized interests for scarce resources. Certain aspects of theories of technological change articulated in recent years fly directly in the face of this model. Social scientists have become aware of the fact that many of the changes that affect people are truly "unanticipated" or "unforeseen." Very often these technology-associated alterations take place with remarkable speed and are, in some cases, "irreversible." What this means is that possible interest groups which could form around an issue concerning the effects of technological change will in many cases simply come too late. The effects will have already taken hold. The business of voicing one's interest at this point can have little significance other than as a plea for redistribution of benefits or for reparations on the injuries suffered. A new class of "losers" is born into the political system.[110]

Illustrations of how this occurs can be found in the recent outcries about air, water, and noise pollution or the loss of privacy brought by the spread of sophisticated information technology. In cases of this sort, a person or group often does not know that his interests are in jeopardy until the harvest has been reaped. In a world of rapid technological change, there may well arise a new mode of politics to stand with such time-honored democratic practices as pork barrel and log rolling— namely, "barn door closure," referring to the normal course of action after the horses have departed. Aware of this predicament and of the problems it poses for pluralist theory and practice, political scientists have begun to work out a new compromise. Their prescription for the politics of technological drift combines the resources of an improved social science and a resuscitated pluralism. Very briefly, the signs pointing to an eventual cure are these: empiricism plus renewed diligence.

The empiricist phase of the solution would strive to obtain that which we now lack: adequate knowledge of the ways in which technology acts to change the human environment. Programs of research would be established to study the various "impacts" of new technologies and to provide citizens and policy makers with advance information "intelligence" concerning possible alternative futures. With this information at its disposal, the society would, presumably, be better able to steer a wise course, bringing the "externalities" and "side effects" under conscious control.

The second aspect of the plan would seek to mobilize latent constituencies in the populace by informing them of their real but unrealized interests in impending technological changes. A report written by students of technology, including Harvey Brooks, Melvin Kranzberg, Herbert Simon, Gerard Piel, and Louis H. Mayo, urged that government agencies seek to encourage "a wide diffusion of deeper understanding about technology and deeper concern about its implications" and that they try "to induce as many different elements of society as possible to involve themselves in broadened [technology] assessment efforts."[111] The report noted that "the habitual tendency to presume technological trends harmless until proven otherwise can be explained by the absence of any group or institution whose function it is to marshal the strongest possible case against a particular trend."[112] New constituencies, therefore, should be created to take care of this unfortunate lack.

One cannot help but admire the ever-recurrent ability of liberal thought to perform marvelous patch jobs to remedy its own flaws. Yes, the engine of change is running amok. But with more data, new studies, more funding, a renewed awareness, an alarm clock under the pillow, and a few minor adjustments here and there on the Madisonian mechanism, we can return to normal. The consumer movement, ecology action groups, and technology "assessors" are agreed that this is the proper course.

For reasons discussed earlier, however, I am not persuaded that the deficiencies in the program can be taken care of through increased doses of research data or empirical theory. The fact that the discussion now wavers erratically between analyses of causal impact and enthusiastic affirmations of free will is an indication that there is something defective in our view of things, particularly in our notions of "choice" and "control." The problem of "unintended consequences," an idea that now threatens to become doctrine, deserves further consideration. What are the characteristics of these consequences and also of the intentions that, we suppose, precede them?

If one examines what are now talked about as "unintended consequences" of technological innovation, one finds that they can be grouped according to the manner of the effects and the medium in which they occur. Much of the discussion centers on truly causal results in purely physical systems. Attention is given to the unanticipated effects of pesticides and herbicides in the biosphere, the unforeseen side effects of drugs and food additives on the human body, and a wide variety of similar maladies involving the long-range effects of fertilizers, detergents, heat pollution, particles of air-carried asbestos, and the like, many of which have come as a genuine surprise to the society. Effects of a strictly causal sort may be either simple or complex. In other words, the "consequence" at hand may have occurred through an uncomplicated, one-directional process which is relatively easy to trace. A particular chemical—DDT, mercury, or some other—is dumped into a brook and eventually harms wildlife downstream. Other sorts of effects, however, come as the result of complex, cumulative processes that are more difficult to observe or pin down. One role of the environmental movement, insofar as it is anything more than a revived anti-litter campaign, has been to advertise the biological disasters, real or potential, that arise from the complex, cumulative results of technological practices.

By way of contrast, there are many kinds of interesting "unintended consequences" that have their effect on individual human beings or within social systems. In cases of this sort it sometimes makes sense to use the ideas of simple or complex causation to describe the changes at hand. We often speak of automation causing unemployment, the introduction of the automobile causing changes in the living patterns of the American family, and so forth. In many instances the ultimate effects we notice were not in any real sense chosen either in the original innovation or in the course of subsequent "use." It is important to note, however, that the language of causation cannot be used exclusively here. Many of the most interesting "unintended" effects are those involving the desires and choices of conscious human beings. Here we are entitled to employ, or indeed must employ, the concepts of intentionality, use, and moral consequence. A new technology, particularly a new technique or apparatus, opens a wide range of practical possibilities. It is ambiguous as to use (which, of course, includes misuse). In many cases the directions of its social application are not known in advance. Modern history is filled with examples of inventions whose practical implications were not know to the inventor. Pascal, Leibniz, and Babbage had no inkling of the future uses of their calculating machines. Poulsen could not have foreseen the range of utility of his magnetic tape recorder. Philo T. Farnsworth had only the most limited sense of the social meaning of his television tube at the time of its development. The early days of each new technology are filled with a sense of pregnant possibilities, along with a profound uncertainty about the eventual outcome. And there is a sense in which we can say that a technical novelty has a life of its own as it finds its way into the complex sphere of social practice.

To read the current literature on ecology or the social impact of technological innovation, one might suppose that uncertainty, unpredictability, and uncontrollability of the consequences of action are situations entirely unique to the present age. The prevailing opinion appears to be that in simpler times, in times that moved more slowly along less complicated paths, such concerns were unknown. But if one turns to the history of Western political thought, one discovers that these same conditions have been known and commented upon since the beginning. Philosophers have identified uncertainty, unpredictability, and uncontrollability as characteristic of all action—factors that political men must take into account before they intervene in the affairs of the world. An actor does not and in fact cannot know the full range of consequences of his deeds in advance. In one of the earliest pieces of political writing available to us, a poem by Solon of Athens, we find the warning:

Danger, for all, lies in all action, and there is no
 telling which way the end will be after a thing
 is begun.
One may be trying to do well and, through failure of
 foresight may fall into the curse of great disaster, while one
who acts badly may find God gives him all that he
 asked for. . . .[113]

Sentiments of this sort are neither rare nor of small importance. They have recurred frequently in the history of Western thought, and the doctrine of "unintended consequences" is merely their most recent version. Taken together, ideas of this sort amount to a countertradition to the dream of mastery. In this tradition, the world is not something that can be manipulated or managed with any great assurance. The urge to control must inevitably meet with frustration or defeat. The more extreme statements of this kind maintain that the universe is ultimately governed by chance, chaos, fate, fortune, or some other principle that through its workings makes all of human effort an exercise in prideful futility. In Heraclitus and Empedocles, for example, the world is one in which strife constantly disrupts and overturns the works of mortals. According to Heraclitus, "Fire will come and lay hold to all things." "The fairest order is but a heap of garbage emptied out at random." In the thought of Epictetus and the other Stoic philosophers we discover that all action must be informed by an awareness of the fact that there are many things in the world simply beyond our power to control. Other than what he may decide to do with his own soul, each man has little mastery over his fate. In Machiavelli the evanescence of worldly circumstances is announced in the opposition of *fortuna* and *virtù*. Fortuna is like "an impetuous river that when turbulent, inundates the plains, casts down trees and buildings, removes earth from this side and places it on the other; every one flees before it, and everything yields to its fury without being able to oppose it."[114] While Machiavelli believed that the *virtù* of a bold and intelligent actor could direct the forces unleashed by *fortuna* to his own purposes, his model of action is distinctly non-Baconian. The world is not one of predictable regularities and passive variables subject to simple control; it is ultimately one of extreme caprice in which an actor must employ the most subtle of intuitions and artful measures to keep from being swept under. Even then he can never be certain.

The condition that lies at the root of such uncertainty, unpredictability, and uncontrollability in the circumstances of action is the complex interconnectedness of the world. Again, the contemporary literature implies that this situation is a revolutionary discovery by ecologists and technological systems analysts. In point of fact, this insight has long been fundamental to the antimastery tradition in Western philosophy. Marcus Aurelius saw the situation clearly enough to observe, "All things are implicated with one another, and the bond is holy; and there is hardly anything unconnected with another thing."[115] The state of interconnectedness in both nature and human society has long been appreciated as the primary source of risk in action of all sorts, for a disturbance of the complex web of relationships begins a process over which the actor has only limited control. In *Search for a Method,* Jean-Paul Sartre observes that "the consequences of our acts always end up by escaping us, since every concerted enterprise, as soon as it is realized, enters into relations with the entire universe, and since this infinite multiplicity of relations goes beyond our intention."[116]

An interesting recent analysis of this state of affairs, particularly as it pertains to modern science and technology, is contained in the writings of Hannah Arendt. "Action," she points out, "no matter what its specific content, always establishes relationships and therefore has an inherent tendency to force open all limitations and cut across all boundaries."[117] "To do and to suffer are like opposites of the same coin, and the story that an act starts is composed of its consequent deeds and sufferings. These consequences are boundless, because action, though it may proceed from nowhere, so to speak, acts into a medium where every reaction becomes a chain reaction and where every process is the cause of new processes."[118] To some extent, individuals are able to set limitations which protect them from the condition of boundlessness. But they cannot manage the second important characteristic of action: "its lack of adequate foresight, but also that the *meaning* of the deed and its consequences can only be clear when the chain reaction which springs from it is complete."

For reasons not relevant to the present discussion, Arendt holds that political action has, by and large, vanished in the modern age. However, there is one variety of action—the works of science—that still remains. "The capacity for action, at least in the sense of the releasing of processes is still with us, although it has become the exclusive prerogative of the scientists, who have enlarged the realm of human affairs to the point of extinguishing the time-honored protective dividing line between nature and the human world."[119] By "the releasing of processes" Arendt means the starting of chain reactions "whose outcome remains uncertain and unpredictable whether they are let loose in the human or the natural realm."[120] The true actors of our time are the scientists, even though they act indirectly with regard to society and politics and have no knowledge of the political tradition which could help them understand the meaning and limitations of their deeds.

The uncertainty and uncontrollability of the outcomes of action stand as a major problem for all technological planning. If one does not know the full range of results that can spring from an innovation, then the idea of technical rationality—the accommodation of means to ends—becomes entirely problematic. The means are much more productive than our limited intentions for them require. They accomplish results that were neither anticipated nor chosen and accomplish them just as surely as if they had been deliberate goals. Nietzsche recognized this situation as a fundamental weakness in the philosophy that guided the scientific and industrial development of his time, utilitarianism. "The value of an action must be judged by its consequences—say the Utilitarians—; to judge it by its origins implies an impossibility, namely that of *knowing* its origins." "But does one know its consequences? For five steps ahead, perhaps. Who can say what an action will stimulate, excite, provoke? As a stimulus? Perhaps as a spark to touch off an explosion? —The Utilitarians are naive—And in any case we must first *know what* is useful: here too they look only five steps ahead—They have no conception of the grand economy, which cannot do without evil."[121] Few would want to agree with Nietzsche's view that attention to the consequences is a total absurdity. At the same time, it is clear that the apparent rationality of our plans, designs, and calculations often collapses in the broader set of relationships and events. A most notorious example is the present environment of the American city, which is much like a colossal sociological truck farm in which one rational plan and technological innovation after another has borne unexpected and unwanted fruit. Under the label of "systems approach" many social scientists have now begun to recognize the "grand economy," which, as Nietzsche pointed out, "cannot do without evil."

The condition noted by the antimastery tradition is one which presumably cannot be overcome. In this way of looking at things, there will always be some consequences of any human intervention in the world that cannot be anticipated or controlled. "The reason why we are never able to foretell with certainty the outcome and end of any action," Arendt observes, "is simply that action has no end. The process of a single deed can quite literally endure throughout time until mankind itself has come to an end."[122] It is clear, nonetheless, that many of the occurrences and conditions we now speak of as "unintended consequences" of technological innovation—pollution, drug abuse, unemployment, the congestion of the cities, the generation gap, megalopolis, and various other ills in nature and society—are often things that could very well have been foreseen and avoided. The world is not so complex and fast moving that, for example, the rising decibel count in urban areas could not have been anticipated and limited before it became a hazard to health. But in most cases of this sort, we find ourselves both surprised and impotent—victims of technological drift.[123]

Along with those effects, then, which are absolutely unforeseen and uncontrollable are those which are susceptible to foresight and control but are never limited by either one. To see why they occur in this manner we should consider two more peculiarities of the "unintended consequences" that have popped up in the recent discussion: (1) that they are almost always negative or undesirable effects and (2) that unintended consequences are not *not* intended.

In the main, we are not particularly interested in those side effects of technological innovation that do not injure, disrupt, jeopardize, or destroy. Positive side effects, everyone understands, will take care of themselves and require no special attention. The aura of negativity that surrounds the present discussion does not mean that there are no unanticipated effects of a beneficial sort, but merely that in speaking of "unintended consequences" we wish to refer to something undesirable.

Unintended consequences are not *not* intended. This means that there is seldom anything in the original plan that aimed at preventing them. Until recently, the idea that any precautions should be taken when a new technology entered social practice was virtually unheard of.

Taken together these notions point us toward a fact that has been a part of the tacit knowledge and most basic commitment of Western society for the last two hundred years. Although it is seldom stated explicitly, we have now seen enough to put it into words: *technology is most productive when its ultimate range of results is neither foreseen nor controlled.* To put it differently, technology always does more than we intend; we know this so well that it has actually become part of our intentions. Positive side effects are in fact a latent expectation or desire implicit in any plan for innovation. Negative side effects, similarly, are experienced as necessary evils that we are obligated to endure. Each intention, therefore, contains a concealed "unintention," which is just as much a part of our calculations as the immediate end in view.

Imagine a world in which technologies accomplish only the specific purposes one had in mind in advance and nothing more. It would be a radically constricted world and one totally unlike the world we now inhabit. The simple logic of means and ends, tools and use, is ultimately of little help in understanding what technology has to do with change. It is like noticing *B* follows *A* while forgetting the other twenty-four letters and their myriad of combinations in words and sentences. Technology, in its various manifestations, affects the world by enlarging the scope and power of human activity in general as well as in the specifics. Each new variety of apparatus, technique, or organization expands the sphere of human possibilities to a degree which, in the nature of things, remains uncertain. The laboratory in which the consequences of innovation are probed and analyzed can only be that of history itself.

Our most basic cultural understanding about what technology is and what it does recognizes the state of affairs I am describing. It can be seen, for example, in the fact that utilitarian calculations have never been able to stand by themselves but have always been propped up by embarrassingly foggy notions of "progress." Industrializers and modernizers understood Nietzsche's critique of utilitarianism long before he made it: the consequences cannot be foreseen. All the better, for it is a mistake even to worry about them. To insist that the development of technology be restrained by the desire for rationality or final control is a retrogressive attitude. This was precisely the error of the utopians and the reason why they are now commonly thought to be the fools of history. The utopians wished to know the outcomes in advance. They wanted to manage the development through enlightened, rational planning. What folly! Utopia would be the death of progress. It is not the ticket that scientific technology gives to civilization. Technological development opens doors to a new world, the final shapes of which are unknown. Thus, to have asked in advance what the computer would mean for civilized life or to plan and control precisely the changes it would bring would have been totally antithetical to the way we normally proceed. Those who work on the computing capacity of the machines are seldom interested in the full range of possible applications. Those who tend specific applications have only minimal knowledge of the social effects such applications may bring. In effect, we are committed to following a drift—accumulated unanticipated consequences—given the name *progress.* If the term *determinism* still applies to this pattern of change, it is, paradoxically, a *voluntary determinism,* one which serves us as long as we avoid demanding to know the outcomes too early.

To this day, any suggestion that the forward flow of technological innovation be in any way limited by an idea of rational or humane planning is certain to evoke a harsh response. Such proposals violate a fundamental taboo. Even the writings critical of the side effects of technological change must observe the underlying commitment without question. The writers of a National Academy of Sciences technology assessment report repeatedly insist that "our purpose is not to conceive ways to curb or restrain or otherwise 'fix' technology"[123] and that "the advances of technology have yielded and still yield benefits that, on the whole, vastly outweigh all the injuries they have caused and continue to cause."[125] Expressions of this sort have become a kind of ritual oath, which anyone the least bit critical of technological affairs must administer to oneself before going any further. In the words of one of our more bald-faced proponents of accelerated technological change: "There is no such thing as retrogressing 'a little.' There is no such thing as a 'restrained progress.' You are hearing many voices today that object to an 'unrestricted technology.' A *restricted* technology is a contradiction in terms."[126]

A model of prudence in innovation was outlined in the eighteenth century, albeit in a nontechnological setting, by Jean Jacques Rousseau. It advised sensitivity to the subtle changes that any innovation engenders in social institutions and public morality. But this model is not something the modern age has taken seriously. Instead we accept the role of experimental subjects in a process of minimally controlled change, later looking back upon what we have done to ourselves as a topic of curiosity. Thus, after almost thirty years during which the lives of children have been influenced by watching television, the question, What are the effects of television upon children? is suddenly posed as a legitimate problem for social science. Minerva's owl flies only after the late show.

The Technological Imperative

Our discussion so far leaves one crucial theme untouched, for there is a set of phenomena best described not merely as effects or consequences but as actual requirements of innovation. Here we encounter one of the most persistent problems that appears in reports of autonomous technology: the technological imperative.

The basic conception can be stated as follows: *technologies are structures whose conditions of operation demand the restructuring of their environments.* Again, the point has nothing to do with any occult force. It is simply a matter of specifying what needs to take place before an instrument is in working order.

Operational requirements of this kind may be either purely instrumental or economic.[127] Instrumental conditions include those involved in the establishment and maintenance of the device's own internal structure. In most examples now called "advanced," technologies require other technologies for their successful functioning. The apparatus is useless without connection to a vast array of technique and organization. Organization makes no sense without sophisticated techniques, which in turn derive their raison d'être from the existence of technical hardware. Thus, a chain of reciprocal dependency is established in which the various aspects of a given technical operation overlap and require each other.

Economic requirements, although not in every instance easily distinguishable from the instrumental conditions, are those which concern the provision of resources—energy, materials, labor, information, and so forth. A technological innovation will often create scarcity where none existed before. Things not previously needed for a particular practice now become necessary resources. Before the invention of techniques of heart transplanting, there was no scarcity of hearts; one per person was universally supplied. With the advent of transplants, however, the organ suddenly became a scarce commodity.

To trace the range of instrumental and economic conditions required to put a particular modern technology in working order is often a staggering chore. One need only think of the thousands of requirements that must be met before the automobile can be a functional part of social life—manufacture, repair, fuel supply, highways, to name just a few. Economists and organization theorists refer to the formation of complex webs of this sort as "vertical integration," a state of affairs in which the output of one operation becomes input for the next.[128] Ellul labels the same phenomenon "monism" and "the necessary linking together of techniques."[129]

The technological imperative contains a logic that accounts for much of the way change occurs in modern society. The logic is not that of syllogistic inference. Rather, it is the pragmatic rationale of necessary action. If you desire *X* and if you have chosen the appropriate means to *X*, then you must supply all of the conditions for the means to operate. To put it differently, one must provide not only the means but also *the entire set of means to the means.* A failure to follow the correct line of reasoning in formal logic brings an unhappy outcome: absurd conclusions. Failure to follow the dictates of the technological imperative has an equally severe outcome: a device produces no results (or the wrong ones). For this reason, once the original choice has been made, the action must continue until the whole system of means has reached its proper alignment.

The force of technological imperatives is reinforced by their connection to what are perceived as the necessities of life. Certain technical means stand at the very basis of human survival. Failure to provide for them is to invite discomfort, suffering, or even death. For this reason the technological imperative is much more than a functional requirement. It is also a moral standard, a way of distinguishing the good from the bad, the rational from the irrational, the sane from the insane. It tells us what is necessary for our continued existence and happiness. Any attempt to deny this necessity can only be an expression of malice, stupidity, or madness. If we have chosen to utilize electrical apparatus in many of the basic activities of life, then we absolutely must have all the means necessary for the supply of electricity. We must build as many power plants as are necessary to take care of the need. There are some things we cannot do without.

In technological innovation, therefore, the possibilities widen, but so do the demands. This fact is crucial to the processes of historical change that we identify as industrialization and modernization. David Landes describes the working out of the logic in his description of the industrial revolution:

The invention and diffusion of machinery in the textile manufacture and other industries created a new demand for energy, hence for coal and steam engines; and these engines, and the machines themselves, had a voracious appetite for iron, which called for further coal and power. Steam also made possible the factory city, which used unheard-of quantities of iron (hence coal) in its many-storied mills and its water and sewage systems. . . . And all of these products—iron, textiles, chemicals—depended on large-scale movements of goods on land and on sea, from the sources of the raw materials into the factories and out again to near and distant markets. The opportunity thus created and the possibilities of the new technology combined to produce the railroad and the steamship, which of course added to the demand for iron and fuel while expanding the market for factory products. And so on, in ever widening circles.[130]

This expanding array of technical requirements and economic demands is the basis of one important aspect of technological dynamism. As Landes points out, change is actually a "logical corollary" of technical rationality, "for the appropriation of means to ends that is the essence of rationality implies a process of continuous adaptation."[131] Speaking of the "cumulative, self-sustaining advance" of the industrial revolution he concludes, "In all the diversity of technological improvement, the unity of the movement is apparent: change begat change."[132]

But it would be wrong to conclude that the scope of adaptation is limited solely to instrumental or economic modifications. As the literature of modernization makes clear, the process of change that accompanies technological innovation touches "every dimension of society."[133] All varieties of customs, habits, attitudes, ideas, and social and political institutions are caught up in its flow, altered, and set on a new foundation. Nothing is left untouched. W. W. Rostow's discussion of "The Underlying Process of Modernization" gives a sense of the broad sweep, practical necessity, and moral urgency involved in these adjustments: "Psychologically, men must *transform or adapt* the old culture in ways which make it compatible with modern activities and institutions. The face-to-face relations and warm, powerful family ties of a traditional society *must give way*, in degree, to new, more impersonal systems of evaluation in which men are judged by the way they perform specialized functions in the society. In their links to the nation, to their professional colleagues, to their political parties, to their labor unions, *men must find* a partial alternative for the family, clan, and region. And new hierarchies, based on function, *must come to replace* those rooted in land ownership and tradition [emphasis added]."[134]

A perspective which looks at the way technological change generates imperatives for society is useful in explaining aspects of modern development which an analysis of the "unintended consequences" of technical action may overlook. One of its advantages is that it recognizes how a process of adaptation can and frequently does precede the technological innovations requiring that adaptation. We speak of requirements that are met beforehand as "meeting the preconditions." The environment is modified to make room for the thing or things demanding that modification; in a certain sense the effects antedate the cause. Most of the nations that are now called "underdeveloped" or "modernizing" (teleology anyone?) do in fact proceed in this manner. Their path is that of emulation in which already modernized nations are studied to discover the conditions necessary to build a stable technological order.[135] Then the "preconditions" are put into effect. For Rostow and many other students of modernization, this means that whole cultures must be literally ripped apart and reassembled before the "take-off" of the great airship "modernity" can begin. Except on the very earliest flights, tickets must be purchased in advance.

I do not wish to leave the impression that the logic of such imperatives is something entirely unique to technological affairs. In a broader sense, what is involved here is a very general problem that accompanies any significant life choice of any kind, for one's decision implies a commitment to support the decision; the initial act requires a sequence of actions to sustain what has been done. Anyone who has ever had a child knows this logic full well. Having a technology is very much like having a child except that its conditions are even more routinized and extensive.

In this interpretation, the notion of a technological imperative recognizes a particular state of affairs. It suggests that the very construction of technological systems contains an inherent tendency to establish a complex set of linkages that continues beyond one's original anticipation and that carries a powerful force of social obligation. Under a technological imperative one does not always know in advance the requirements that a new technology carries with it. Even the most clear-sighted choices may be blind ones. A society is literally forced along paths, compelled to make enormous investments, that it did not select and might have sought to avoid.

John Kenneth Galbraith's *The New Industrial State* is one of a number of recent writings that conclude that a society propelled by imperatives of technology is increasingly closed, inertial, inflexible, and isolated from any true conception of human needs. Galbraith argues that operations of the modern industrial system necessarily require specialization, complex organization, an increasing span of time between the beginning of a task and its completion, increased capital commitment, and planning.[136] The rest of his book grapples with the fact that as a result of these requirements, the industrial system becomes increasingly unresponsive to any ends other than its own survival and growth. Ellul, of course, describes the situation in even more radical terms. "Technique," he writes, "cannot be otherwise than totalitarian. It can be truly efficient and scientific only if it absorbs an enormous number of phenomena and brings into play the maximum of data. In order to coordinate and exploit synthetically, technique must be brought to bear on the great masses in every area."[137]

Later I will perform a more thorough dissection of specific forms that the technological imperative takes. The issue will be recast as a central concern of a general conception of the politics of technology. We drop it now, not because it is of minor importance but because it involves questions too large for the present context.

In each of the instances in this chapter I have described ways in which technological change may be said with some justification to be "out of control." Each case combines an objective state of affairs—an actual process in the world—with a predisposition of men in society to allow the changes to continue with little intervention. Together, the process and the disposition create what can be called technological dynamism, a forceful movement in history which continues largely without conscious human guidance. This is not to say that changes in technology and society are never chosen, directed, or controlled. The reader may wish to send me a list of changes that are truly voluntary, and I will gladly respond with a list in agreement. The point is that there are important categories of change that simply do not make sense under the ideas of "chosen" or "voluntary." Again, the circumstance that makes this interesting is that technologies and the results they produce are exactly the kinds of things that we expect to be able to choose and control.

Chapter 3

The Flaw and Its Origins

There is no obvious path to follow to the belief that technological dynamism is an important problem in the modern world. Experiences of many sorts, analyses of various kinds lend plausibility to the idea. In typical cases an observer focuses upon a particular set of historical alterations—in work, the environment, social institutions, or some other sphere—that is in some way disturbing and then considers how the situation might have been different. More often than not the answer involves the need for controls on a local, national, or international level far beyond anything even remotely feasible. A common response to this insight is despair or blind rage at the spectacle of personal and political powerlessness. It is this despair for which the numbing salves of the idea of progress (which made it something of a virtue to endure congested cities and pollution) were once, but are no longer, the perfect cure.

We have already noted some obvious channels open to those able to avoid paralysis and resignation. One can assume the role of a firefighter, select one or more of the areas of life in which technological change looms as a problem, and set one's goals to improve things in that sphere. There remains the opportunity to mount campaigns to modify or block the planning of ill-considered urban highways, to regulate the manufacture and distribution of dangerous insecticides, or to alert the populace to the dangers of propellants in aerosol cans. A new breed of public-interest scientists, engineers, lawyers, and white-collar activists now pursues this demanding vocation. One can only wish them well.

But a therapy that treats only the symptoms leaves the roots of the problem untouched. Intuitively, at least, we understand that the various separate issues that concern us here may have common origins deep in the history of human culture. Roughly the same tendencies that produce the chaotic living conditions of megalopolis are those that create hundreds of superfluous consumer innovations like digital clocks and Teflon-coated razor blades. Roughly the same motives and outlook that encourage greater and greater industrial productivity, rapidly expanding investment in research and development projects, and an unprecedented rate of innovation in all spheres of life are also those mirrored in the "need" for a new manned bomber, the eutrophication of lakes, and the willingness to "solve" serious social problems with a tech-fix. If we are to understand the dispositional element in technological dynamism, it appears that we must go beyond cataloging symptoms to a level of awareness and therapy provided by depth analysis.

A major concern in philosophical writing on technology and culture is to offer diagnoses of exactly this kind. The twentieth century has produced a great body of theory which searches earnestly for deep metaphysical roots, historical origins, and elements of essential unity in the diverse phenomena that characterize technological society. In the pages that follow I will consider some paths which attempts at deep critique have taken, in particular those routes which approach the question of an unlimited, excessive growth of technology. What claims do theories of this sort attempt to make? How well founded are the hypotheses that come from such inquiries? Most important, does one learn anything from deep-seeking analyses that might be helpful in reorienting action in the technological sphere?

Nature and Western Civilization

Perhaps the most common variety of hypothesis in the speculative literature places blame for the excesses of technological civilization on a flaw in the nature of man. Aggressiveness, acquisitiveness, restless desire for conquest and power, or some other allegedly central trait is cited as the root of problems in the modern condition.[1] In many formulations the qualifier "Western" is added to the general category "man" to allow for the obvious fact that recent advances in science and scientific technics have been, by and large, distinctly European and North American projects. This qualification is of no small importance in such arguments, and I shall return to it in a moment.

Statements of the flaw-in-the-human-character thesis can be found at many points in the writing of Lewis Mumford. Long an opponent of the idea that oppressive, large-scale technologies are the exclusive invention of modern times, Mumford finds the source of the technological dilemma in an indelible affliction in motives which has disrupted Western culture since the Egyptians. Man has a choice between authoritarian megatechnics (large-scale systems) and democratic polytechnics (small-scale systems, arts and crafts). The megatechnical obsession seeks to bring everything under control and thereby places all of nature and society in peril. The polytechnical tradition, on the other hand, is inherently moderate. Small-scale tools are used in carefully conceived ways with a minimum of unanticipated side effects. A choice for megatechnics over polytechnics is for Mumford basically that of evil over good. The roots of our present situation are to be found in the prideful, ignorant, power-hungry technical mentality reintroduced to the modern age by Francis Bacon. "It was not technological insight and adroitness," Mumford observes, "but cupidity, power-hunger, overweening pride, and indifference to the future that kept Western peoples from maintaining their own craft traditions and tool-using habits."[2]

Mumford believes that avariciousness and power mania in human beings continually threaten to assume dominance. The human essence is not defined by this trait alone, however. He argues that man is a multifaceted, self-transforming combination of an original biological nature and a "second nature" created over thousands of years.[3] Humans have the potential to become, in effect, many different kinds of creatures. In this respect Mumford finds a lethal link between the attitude modern man takes toward himself and the flaw that is buried deep in his original nature. The modern self-image rests on an insidious myth that man is essentially a tool-making animal. Wedded to the tremendous power of modern invention and discovery, this myth encourages an unrestrained growth of megatechnical monstrosities. Men who believe that their nature is expressed in technological projects can find no sense in the idea that unlimited technical development might lead to excess and aberration. Against the view that humans are essentially tool-making, tool-using creatures, Mumford marshals a mass of evidence from anthropology and history leavened by his own optimistic interpretation. Homo sapiens, he argues, precedes homo faber. Man the mind-maker developed his capacities of consciousness, imagination, and intelligence long before material instruments became a concern. This is shown, Mumford insists, in the record of ritual, symbolism, and noninstrumental creativity which appears at the very earliest stages of human culture.[4]

Barnett Newman, the American abstract expressionist painter, made a similar point in an extraordinary essay published in 1947. "What was the first man, was he a hunter, a tool-maker, a farmer, a worker, a priest, or a politician?" "Undoubtedly," Newman concludes, "the first man was an artist."[5] The crucial turning point in human evolution was not the use of weapons or tools. "Man's hand traced the stick through the mud to make a line before he learned to throw the stick as a javelin."[6] Of course animals had always produced lines inadvertently as they moved through the world. But it was only when one of them drew a line with a stick, looked down, and saw both itself and the universe reflected in that mark that the boundary between animal and human was crossed. "Man's first expression, like his first dream, was an aesthetic one. Speech was a poetic outcry rather than a demand for communication. Original man, shouting his consonants, did so in yells of awe and anger at his tragic state, at his own self-awareness and at his own helplessness before the void."[7]

Affirmations of this kind must remain almost entirely speculative. Evidence from Olduvai Gorge relics, stone-age cave paintings, and other such sources is too fragmentary and ambiguous to permit any final conclusions about man's earliest activities, much less what they tell us about "human nature."[8] As Newman himself points out, paleontology "has entered a realm where the only questions worth discussing are the questions that cannot be proved."[9] Thus, Lewis Mumford's exaggerated stress upon man the mind-maker stems only in part from his scientific interest in the issue. His writings and lectures on the subject are obviously intended as moral correctives for the one-sided prejudice that informs twentieth-century views of "technological man." If the combined wisdom of anthropology and the other social sciences tells us anything, it is that the quest to find a unitary essence or characteristic activity that defines the human being is ultimately futile. What one discovers in the human population is a multiplicity of traits that are emphasized or repressed in various cultural settings. Most certainly, there is plenty of evidence that man in many times and places has lived totally without an urge toward unlimited technological action and insatiable consumption. Faced with claims that the problems of a technological age lie "deep within us," that is, within our structure (presumably our genetic structure) as human beings, it is wise to remain skeptical. In the end, I am inclined to agree with Hannah Arendt's argument that if there is a human essence, it would require a god's-eye view to know it.[10]

One way around the difficulties of a purely naturalistic hypothesis is to look for origins of technological dynamism in the distinctive character of Western civilization. It is the West rather than other cultures of the world that before recent times took upon itself the vast enterprises that characterize modern industrial development. "Our technology has absorbed elements from all over the world, notably China," one student of the subject points out, "yet everywhere today, whether in Japan or in Nigeria, successful technology is Western."[11] The primary feature that appears to distinguish the West in this regard is its commitment to understand, control, and exploit nature. Scientific intelligence probes the structures and processes of natural phenomena so that they can be harnessed to an infinite variety of works devised by technical skill. This, the argument usually goes, is a very ancient tendency in the culture, which has been realized by the success of science and scientific technology over the past three centuries.

But if the impetus to dominate nature is somehow built into our way of life, the question becomes that of locating specific aspects of Western civilization that engender this tendency. Are there institutions, belief systems, or some other permanent aspects of our heritage that induced Western peoples to take up technological activism?

The most common of existing hypotheses on this theme focus upon the Christian religious tradition. Max Weber's *Protestant Ethic and The Spirit of Capitalism* set the tone for such discussions and provided the first significant theory. Contradicting the Marxist notion that social ideas mirror forces and relations of production, Weber argued that the capital-building impulses of the bourgeoisie were secularized versions of a distinctly Calvinist response to uncertainty in the face of God's grace. While the fact of salvation could never be known, its likelihood could be demonstrated in worldly virtue, including success in material enterprise. Channeled into finance, industry, and the rationalization of a broad range of productive activities, this feature of Calvinist ideology played an important part in shaping the social roles of early capitalism.[12]

Weber's thesis has had an important influence upon scholarly research into the social origins of modern science and technology. But for some observers, his explanation fails to delve deep enough in its pursuit of the source of rationalization and the domination of nature. In a widely reprinted article, "The Historical Roots of Our Ecologic Crisis," Lynn White, Jr., speculates on the question of why the powers of scientific technology "to judge by many of the ecologic effects, are out of control."[13] Western ascendance in such matters, he argues, took shape long before the modern scientific and industrial revolutions. Very early in medieval Europe, certainly no later than 1000 B.C., important developments in power machinery and labor-saving devices began to characterize the West. "Not in craftsmanship but in basic technological capacity, the Latin West of the later Middle Ages far outstripped its elaborate, sophisticated, and aesthetically magnificent sister cultures, Byzantium and Islam."[14] Similar progress also occurred in medieval science. To find the wellspring of such activity, White suggests, one need look no further than the beliefs of the early Christian church. Christianity "not only established a dualism of man and nature but also insisted that it is God's will that man exploit nature for his proper ends."[15] Church theology taught that God had created men as transcendent beings and that nature was provided to them as a garden to be ruled. At the same time the faith discarded the conviction of pagan animism that since every flower, tree, and stream had its own guardian spirit, care must be taken not to disturb a natural object without sufficient reason. Christianity brought with it a much more cavalier, even ruthless attitude toward the relationship of man and nature. It is the secularized form of this attitude, White argues, that shapes the Western propensity for unlimited technological action today.

Biblical accounts do avoid the pitfalls of explanations which begin with "human nature." But they have serious shortcomings of their own. In making his case, White is extremely selective in choosing those features of Christian belief that he upholds as formative. His account stresses the importance of the Old Testament story of creation. "Man named all the animals," White reminds us, "thus establishing his dominance over them. God planned all of this explicitly for man's benefit and rule: no item in the physical creation had any purpose save to serve man's purposes."[16] While White's article was never intended to be a complex piece of biblical exegesis, its view stands or falls on the validity of the idea that at the heart of medieval Christianity stands a doctrine which encourages man's mastery and rule over nature. True, Genesis does say that God directed man and woman to "be fruitful and multiply, and replenish the earth, and subdue it: and have dominion over the fish of the sea, and over the fowl of the air, and over every living thing that moveth upon the earth."[17] But what would happen if other scriptural passages were taken as the touchstone? Christ's Sermon on the Mount, to cite only one possible example, tells us: "Blessed *are* the meek: for they shall inherit the earth."[18] It is not easy to derive an impulse to dominate nature from that. Indeed there are many elements of Christianity that enjoin moderation, passivity, and ascetic denial in approaching the world of material concerns. Many of these elements were important parts of the teaching of the medieval church.[19] What, if anything, entitles us to say that Christian theology is the ultimate ground of the tendency toward unfettered technological activism?

An eloquent critique of White's theory marks the starting point for John Passmore's reflections in *Man's Responsibility for Nature.* Passmore shares White's concern for environmental degradation and sets out to find "what the West has to jettison, and what to retain, if it is to have any prospect of solving the problems which confront it."[20] He agrees that much of Christian theology has helped justify the arrogant, aggressive stance of the West toward nature. But he finds no substance in the view that this malady has its source in what Western culture has learned from Genesis. Calling upon evidence from both scripture and biblical history, Passmore concludes that the Hebrews recognized no fundamental gap between man and nature. Nature was created first "and God saw that *it was* good."[21] Indeed, when man was created he was given dominion over the earth and its creatures. But, Passmore points out, the Old Testament "is uncompromisingly theocentric: nature, on its view, exists not for man's sake but for the greater glory of God."[22] It is a mistake, therefore, to trace the arrogance of Christian theology toward nature back to the beliefs of the Hebrews. Passmore goes on to argue that its true source is none other than Greek philosophy as it influenced Christian ideas. Aristotle's *Politics* expresses the idea that animals were created for the use of men. The Stoics later carried this view "to extraordinary lengths," seeing in the fact of human rationality a clear warrant for man's mastery over all lesser beings. By its very design, nature had been created to serve human needs. This doctrine of the Greek and Roman Stoics was embraced by Christian thinkers, most notably Origen in the third century A.D. As Passmore points out, however, the Stoic notion by itself leads to quietism. "It is only when it is coupled with a Pelagian, humanistic, attitude to man, which sees him not as essentially corrupt but as having the duty to create, by his own efforts, a second nature—identified, in the Christian West, with a second Garden of Eden—that it can either provoke or be used to justify a scientific-technological revolution."[23]

Passmore's treatment of the flaw in Christian theology represents an improvement over the simplicity of White's account. He is aware that there are many subtleties of doctrine to be taken into account, that the faith does not speak with one voice on the matter, and that one figure of some importance, Jesus of Nazareth, seems to have had no truck whatsoever with the idea of man as despot. Nevertheless, other than proving that even the ecologist can quote scripture, Passmore's efforts to achieve a decisive resolution are unsatisfactory. The reason is that one is never completely clear about what is being asserted. This is a problem endemic to writings on the domination of nature, not merely those which focus on religious origins, for whether the source is found in Genesis, the Stoics, or, as is often asserted in secularized versions, Francis Bacon and René Descartes, the suggestion is made that a particular strand of theology or philosophy has powerfully shaped Western scientific and technological practice. But the extent of the argument remains unclear. Are we to believe that the idea of the domination of nature became a central part of Western education, that it was taught by the church or in schools and universities? Did the idea somehow enter the everyday world view, influencing the activities of a particular social class or everyone in the culture? Whatever answers to such questions might be forthcoming, explanations like those given by Passmore fail to forge the crucial gap between narrow doctrine and broad social practice. We are asked, in effect, to make exorbitant inferences and fill in steps that the theory itself leaves out. In Passmore's case this results in asking us to believe that much more attention was paid to the Stoics in Western history than even the wildest overestimation of their influence would merit.

Much of the attraction of the domination of nature thesis, I suspect, lies in the seeming profundity of its dramatic images. There is something oddly appealing in the idea that the troubles of our technological age stem from a perverse streak in the very identity of Western man. The vogue of the notion has made it an important theme in mid-twentieth-century philosophy, addressed seriously by certain schools of neo-Marxian criticism and phenomenology. During the 1940s Max Horkheimer and Theodor Adorno argued for the centrality of the impulse to dominate nature as an alternative to Marx's original explanation of the root of the ills of industrial society. A theory that took class struggles in the context of capitalist accumulation as its starting point was replaced by a deeper seeking vision of the flaw in Western culture. Horkheimer's *Eclipse of Reason*, for example, sets out to account for certain crucial tendencies of thought—subjectivism, positivism, pragmatism, and instrumentalism—which, he maintains, characterize the pathological direction Western reason has taken in the modern age. "Having given up autonomy, reason has become an instrument. In the formalistic aspect of subjective reason, stressed by positivism, its unrelatedness to objective content is emphasized; in its instrumental aspect, stressed by pragmatism, its surrender to heteronomous contents is emphasized. Reason has become completely harnessed to social process."[24] Such tendencies of thought, however, must be analyzed to their ultimate source. "If one were to speak of a disease of reason affecting reason, this disease should be understood not as having stricken reason at some historical moment, but as being inseparable from the nature of reason in civilization as we have known it so far. This disease of reason is that reason was born from man's urge to dominate nature, and the 'recovery' depends on insight into the nature of the original disease, not on a cure of the latest symptoms."[25] Horkheimer observes that this mentality can be traced back as far as the Book of Genesis. In *Dialectic of Enlightenment* Horkheimer and Adorno find the same tendency fully present in the thought and motives of Homeric Greece.[26] The location of origins, however, is less important to the two philosophers than their belief that the historical development of enlightenment, the very progress of human reason, is inextricably bound up with projects of domination.[27]

Indeed, any hypothesis that seeks to identify contemporary tendencies toward uncontrolled technological expansion with specific historical origins takes on a difficult task, for what does it mean to locate the responsibility with the ancient Greeks? The Hebrews? The medieval Christians? And what do the alleged spiritual or intellectual roots have to do with modern practice? Do they justify it? Motivate it? Encourage it? As to their ability to provide an intelligible explanation on these counts, the existing theories are simply incomplete.

Beyond such shortcomings, however, lies the even more crucial question of whether, issues of methodology aside, the thesis is even roughly true. Is the conquest of nature, as asserted in such writings, actually a product of a despotic Western urge to dominate? Or is it more accurately seen as a by-product of activities that flow from a more complex variety of ideas and intentions? Certainly, scientific and technical works of the modern age have brought vast areas of nature into the domain of human manipulation and control. True also, certain influential thinkers, most notably Francis Bacon, have proposed the goal of bringing nature under man's governance as a noble motive for science. But neither of these facts justifies the conclusion that all of science and technology is driven by conceptions of and a mania for domination. If one examines the actual activities of scientists, one finds them working to make sense of masses of data, to formulate critical experiments in search of universal laws, and to make progress on what are often very limited topics of research. Engineers, similarly, spend most of their time searching for solutions to very specific, often mundane, practical problems. Indeed, the combined result of these activities constitutes an increasingly powerful mastery of the material world. But it is both superfluous and unfair to suppose that an active desire for domination lies at the heart of every aspect of scientific and technical work.

The idea that Western civilization stands alone subject to this tendency is, it seems to me, also highly suspect. Interference with natural processes and manipulation of the material environment has been practiced in some form by every culture. Human beings must control nature, if only to a minor extent, in order to survive. Indeed, Europeans and North Americans have succeeded in devising astoundingly effective means for doing so. Their tools and techniques are now exported, with sometimes unfortunate consequences, to non-Western cultures. But to conclude from such evidence that the West is unique in its capacity for destructive interference with nature is a mistake. In recent years the American Indians and certain Oriental cultures have been upheld as models for man's harmony with the environment. Archeological evidence, however, indicates that the Anasazis of the American Southwest despoiled forests and land under pressures of a growing population. Those who look to the East should remember the deforestation of northern China already complete by the thirteenth century A.D. Criticisms of the excesses in one's own culture are not aided by the invention of mythical contrasts.

I shall return to these issues later as we consider the application of deep critique to the search for a "new ethic." Now as a contrast to what we have just seen and because it has intrinsic merits of its own, I want to turn to Jacques Ellul's thoughts on the roots of technological dynamism. Ellul touches upon many familiar themes, but he gives them a novel twist.

The Victory of Technique

Ellul's *The Technological Society* is sometimes understood to be a theory of technological determinism. In the strict sense, however, that interpretation is not correct. Unlike Marxist scholars, Ellul does not read history to see how the forces of production have shaped social configurations and the consciousness of men since the very dawn of civilization. On the contrary, he holds that the influence of la technique as a determining force appeared only in relatively recent times. Again and again he stresses that "there is no common denominator between the technique of today and that of yesterday. Today we are dealing with an utterly different phenomenon."[28] What he argues is that in the course of modern history, circumstances arose in which technology finally became the most influential element in civilized life. This development, he asserts, took root in the eighteenth century, matured in the nineteenth century, and bore its full fruit only in our time. "I do not maintain," Ellul cautions, "that the individual is more determined today than he has been in the past; rather, that he is differently determined."[29] "We are conditioned by something new: technological civilization."[30]

Ellul holds that in past civilizations, technique was limited in its sphere of action. It was a purely local phenomenon, strictly contained within the framework of particular societies and was merely one aspect of life in those societies along with religion, art, politics, play, and so forth. "Technique functioned only at certain precise and well-defined times; this was the case in all societies before our own. Technique was not part of man's occupation nor a subject for preoccupation," he observes.[31] In traditional society there was little tendency for techniques to develop rapidly or to migrate across cultural lines. "Technique was unable to spread from one social group to another except when the two were in the same stage of evolution and except when civilizations were of the same type."[32] Above all, the limited techniques of the past allowed the possibility of human choice as to the role technique would play and therefore permitted a wide diversity of cultural forms.

Ellul's survey of world history draws attention to two major examples in which technique was deliberately and successfully limited. In both cases he finds that a religious or moral conviction in the culture prevented an unrestrained development of technical means. The ancient Greeks "were suspicious of technical activity because it represented an aspect of brute force and implied a want of moderation."[33] Their culture was based on notions of harmony, balance, and self-control, and they recognized that technique and its power might force open these limits and destroy the balance that sustained them. "The rejection of technique was a deliberate, positive activity involving self-mastery, recognition of destiny, and the application of a given conception of life. Only the most modest techniques were permitted—those which would respond directly to material needs in such a way that these needs did not get the upper hand."[34] Ellul is fluent in his praise for the Greeks on this matter and is careful to distinguish their response from ignorant technophobia. "This feeling on the part of the Greeks was not a reflection of a primitive man's fear in the face of something he does not understand. . . . Rather, it was the result, perfectly mastered and measured, of a certain conception of life. It represented an apex of civilization and intelligence."[35]

An even more thorough restraint of technique, in his view, took place in Christian civilization of medieval Europe. During the thousand-year period from the fourth to the fourteenth century, technique progressed little at all, and many techniques known to the ancients were lost. The Christians were antiworldly and therefore antitechnical. With the decline of Rome they abandoned the practical operations of that civilization and began a culture marked by the "nearly total absence of technique."[36] Ellul disagrees with White and others that the Christian secularization of nature led to a desire to exploit the material world.[37]

On the contrary, the Christians were suspicious of all practical activity and condemned "everything that represented the earthly city, which was consecrated to Satan and opposed to the City of God."[38] In the later Middle Ages, this stringent antiworldliness lost some of its hold, but another aspect of Christianity remained as a limitation on the growth of technique. The Christian was a severe moralist who insisted that all human activity must be judged according to the standards of good and evil set forth in the word of God. "Technical activity did not escape Christian moral judgment. The question 'Is it righteous?' was asked of every attempt to change modes of production or of organization. That something might be useful or profitable to men did not make it right and just. It had to fit a precise conception of justice before God. When an element of technique appeared to be righteous from *every* point of view, it was adopted, but even then with excessive caution."[39]

Thus, Ellul offers a conception of technology in Western history that contradicts Passmore's treatment of the Greeks on the one hand, and White's conclusions about Christianity on the other. And although it is at least as plausible as either of those accounts, its accuracy is open to question. Much of modern scholarship has tended to reverse the judgment that Greece and medieval Europe were poorly developed in technics. Lewis Mumford has argued that the Greeks were in fact resourceful inventors and engineers and that the true roots of the industrial revolution must be traced back to the technical innovations of the Middle Ages.[40] But Ellul is not speaking of inventions so much as the technical orientation of a whole culture. On balance, it seems to me correct to say that ancient Greece, as represented by Athens and Sparta, and European Christian civilization, as represented by the Church and feudalism, were founded on ideals of social life, personal virtue, and public conduct which at the very least did not greatly encourage technical innovation. One can argue, in fact, that the eventual decline of these civilizations came when the balance of elements sustaining them was disrupted by (among other things) the advent of new techniques. But to go a step further and conclude with Ellul that the highest ideals and virtues successfully discouraged the development of technique is to carry the case too far.

Indeed, it was against the modernizers and technical innovators of Athens and in favor of the life of harmony, justice, and virtue that Plato wrote. Insofar as the *Republic* is a technical treatise, it is a vision of political artifice directly opposed to the new techniques of the time —especially those of trade, finance, and warfare—which he believed would eventually tear the polis apart. In the *Gorgias* he observes. "With no regard for self-control or justice they stuffed our state with harbors and docks and walls and tribute money and all such nonsense; so when this presumed attack of illness finally comes, they will blame the advisors who happen to be about at the time, while praising to the skies Themistocles and Cimon and Pericles, though they were the true author of the trouble."[41]

But protests like those of Plato and his followers were futile in changing the direction that Greece had long since taken. Ellul may be right in identifying the highest ideals of Greek and Christian culture as antitechnical, but he is wrong in finding them a powerful counterforce to technical progress. Actual examples of the suppression of technique are found in much more mundane settings. The guilds and corporations of medieval Europe tended to fear the disruptive effects of technical innovation and often sought to prevent the spread of new developments. In 1578 the Council of Nuremberg destroyed the lathe of craftsman Hans Spaichl because Spaichl had arranged to sell his device to a goldsmith.[42]

When and under what conditions were the cultural limitations on technique which Ellul describes finally removed? His analysis tries to sort through the events leading up to the decisive transformation and distinguish them from the occurrence itself. Although the world of total technology was forming during the Renaissance and scientific revolution, neither the sixteenth nor the seventeenth century was capable of bringing the movement to victory. There were still a few important barriers left, especially the humanism of the period that asserted the supremacy of men over means. "This humanism," Ellul concludes, "did not allow techniques to grow."[43] "Society was at a cross roads. More and more the need was felt to create new means; even the structure these must take was clearly perceived. But the framework of society, the ideas in currency, the intellectual positions of the day were not favorable to their realization. It was necessary to employ technical means in a framework foreign to them."[44]

The lifting of restraints and the explosion of technical innovation came at last in the middle and late eighteenth century. In Ellul's story, this was the time of the Fall for modern man. He explains the birth of technological society as a convergence of a number of historical conditions. Each of these conditions by itself, he insists, would not have been sufficient. But together they were absolutely effective in creating a new world-historical phenomenon.

The first condition was the fruition of a long technical experience. Here he agrees with Mumford that what occurred during this time was a synthesis of previously separate inventions into a "technical complex" or "ensemble," as evident, for example, in the technology of the industrial revolution or of Napoleon's army. A second element was the growth of population, which "entails a growth of needs which cannot be satisfied except by technical development."[45] Also required were certain economic conditions, namely that the economy be both stable and at the same time highly flexible so that new inventions could be absorbed into economic life and further inventions and innovations stimulated. The economic practices of the European bourgeoisie at that time provided exactly this kind of economic environment. Beyond this, Ellul describes a situation that he holds to be most important of all: the plasticity of the social milieu. This, he believes, included two significant aspects: (1) the disappearance of the taboos inherited from Christianity and traditional society that discouraged any tampering with the sacred or natural order of things; and (2) a disintegration of natural social groups that released men into the world as atomized individuals. The culmination of these conditions came in the appearance of what Ellul terms "a clear technical intention" throughout society. Men began to see the usefulness of technique in every area of life and oriented all of their activities toward it.

Ellul adds to this list several other items: the progress of scientific investigation and discovery, the rise of naturalism and utilitarianism in philosophy, and a sweeping change in human attitudes, including a generally more sanguine view of life's possibilities. The two most important parts of his argument, however, are those which concern social plasticity and "the technical intention," for what took place, he believes, was the reverse of what happened at the collapse of Rome. At that time a whole civilization flew apart, and men founded a new one that abandoned worldliness and technique. In the eighteenth century, the hold of the medieval order over the lives of men collapsed altogether, thus leaving the way clear for new modes of action based on technique. In neither case did the shift occur overnight. But there was a critical point in history beyond which there was no return. In this regard *The Technological Society* is similar in its historical position to *The City of God* and in a perverse way attempts to serve the same function. Where Saint Augustine described and justified the formation of a new society dedicated to God, Ellul stands as an ironic devil's advocate giving an account of man's return to Satan's kingdom.[46]

Speaking of the disintegration of "natural social groups" Ellul refers to the cohesive, authoritative structures of collective interest of such institutions as the peasant communes, the guilds, the parliaments, the universities, and hospitals in medieval society. During the French Revolution and the revolutionary ferment of Europe in the late eighteenth and early nineteenth centuries, these groups were attacked and dismantled in the name of the rights of the individual. With these institutions destroyed, men at all social levels were left as rootless, atomized units ripe for reorganization. "The society produced was perfectly malleable and remarkably flexible from both the intellectual and the material points of view. The technical phenomenon had its most favorable environment since the beginning of history."[47] Individuals responded to the chaos and fragmentation by seizing hold of the one reliable thing available at the time: la technique. It was on this thread that they began to reestablish their identity, security, and fortune.

It is important to notice precisely what Ellul means by la technique in this context. His perspective, which is typically French, finds the essence of this transition in reason, more specifically, reason made active in the world. Looking at the Enlightenment and the French Revolution, Ellul finds that both of these were masks for a much more fundamental and consequential movement: the technical revolution. If one looks at the history of the times and asks, What from that period was truly lasting? one finds that it was above all a rationalization and systematization of all activity. "This systematization, unification, and clarification was applied to everything—it resulted not only in the establishment of budgetary rules and in fiscal organization, but in the systematization of weights and measures and the planning of roads. All this represented technique at work."[48] At the very height of chaos, a new order, one based on reason and artifice, began to take form. "From this point of view," Ellul concludes, "it might be said that technique is the translation into action of man's concern to master things by means of reason, to account for what is subconscious, make quantitive what is qualitative, make clear and precise the outlines of nature, take hold of chaos and put order into it."[49]

Ellul argues that the formation of the technological society was not accomplished through a single rational plan or by the systematization of life from any central source. Instead, it was the result of radically decentralized action based on "special interest." By "special" interest, he does not mean "private." The point is that the development and application of techniques, whether by the state or by private concerns, proceeded according to the most narrow of motives and decision criteria. "Will it be effective in this particular instance?" was the only question asked. Contrary to some interpretations of his work, Ellul does not argue that technique does not serve human needs. In his eyes the secret of technique is that it serves virtually *all* such needs and serves them well. One at a time.

An important consequence of the bringing together of special interests and technique in this manner is that any concern for the whole of society vanished altogether. In the early history of the technological society, it would have seemed nonsensical for anyone to have been concerned with the unintended consequences of atomized technical action for the world at large. Within the limits of the "technical intention," such things were simply not at stake. Adam Smith's mention of "the invisible hand" with reference to the welfare of society as a whole briefly recognizes this problem. But the "hand" could merely wave goodbye, as if from a rapidly passing train.

For Ellul, the most active carriers and first beneficiaries of the technical movement were the bourgeoisie. The very identity of this class, he believes, depends on this attachment. A bourgeois is above all else someone who has mastered technique. The bourgeoisie first developed financial and commercial techniques and went on to originate the factory system, the rational administration of the state, technical schools, and so forth. They were truly people suited to their times—marginal, atomized individuals, obsessed by special interest, fascinated by science, and convinced that reason must be made active in every corner of social and material existence. As in the analysis of Karl Marx, the bourgeoisie is midwife to a new society undergoing pangs of birth. Ellul admits that he has learned much from Marx and that he uses Marx's "method of interpretation" and "way of 'seeing' the political, economic and social problems."[50] But in this case, as in many others, he insists that Marx was wrong. In seeing the rise of technological society solely in terms of capitalism, Marx mistook the part for the whole. He focused only upon techniques involved in one significant sphere of bourgeois activity. By seizing upon economic issues alone, and by convincing the proletariat that economic maladies were the source of all possible problems, Marx effectively cut himself and the proletariat off from the true realities of the eighteenth and nineteenth centuries—the total reorganization of life on a narrowly conceived rational basis. It was this process that eventually enveloped and absorbed the proletariat, thus obliterating the last vital source of opposition and alternatives. Marx, Ellul clearly states, was the unwitting servant of this movement, seduced by his own love of technology.[51]

Ellul's conception of the victory of la technique is similar in its structure to Louis Harz's argument in *The Liberal Tradition in America.*[52] Both hold that a certain mode of thought and action, a particular way of defining problems and responding to them, was adopted by a society and then became the dominant pattern that governed universally from that time forward. This type of activity—in Hartz liberal politics, in Ellul technical action—becomes both a conscious and unconscious response to whatever situation arises. This response pattern strongly and automatically repulses any alternative mode of activity. By its own power over the ideas and behavior of men, it neutralizes the alternatives by making them seem unnatural, impractical, or simply impossible. Hartz suggests that "the liberalism of American life, by erecting a set of hidden traps and false facades, confounded not one but all groups who live within it."[53] This way of life frustrated all attempts at true conservatism, socialism, romanticism, and non-Lockean democracy. In much the same way *The Technological Society* points to conceptions of social life eclipsed by the West's growing obsession with technical solutions. Institutions based upon age-old tradition were obvious casualties. The ingenious alternatives for industrial civilization posed by the utopians and anarchists met a similarly unhappy fate.

There were, of course, many voices which spoke in protest of the great transformation Ellul describes or at least suggested that it proceed in a different direction. Early on, Jean Jacques Rousseau criticized the path of progress based on the development of "arts and science." Such supposed improvement was, he argued, actually a form of decadence, which took the human race even further from its essential nature and from the social conditions required to sustain true virtue in the individual. Rousseau's dream for the modern world was one that envisioned the creation of relatively small, self-sufficient communities of free and equal citizens in which considerations of public good would reign supreme. Needless to say, this was not a dream destined to be realized in historical practice. Others who form a minor tradition counter to the rise of technological society include Robert Owen, Charles Fourier, Bakunin, Proudhon, Kropotkin, Ned Ludd and his "army," William Morris, John Ruskin, the aesthetic movement, the dadaists, surrealists, situationists, and a host of other noble losers. In some cases their plans and protests simply lost out in the marketplace of ideas. But it was also true that some were vanquished by force. Speaking in Parliament in 1812 on the matter of the antiframe riots, Lord Byron inquired, "Can you commit a whole country to their own prisons? Will you erect a gibbet in every field, and hang up men like scarecrows? or will you proceed . . . by decimation? depopulate and lay waste all around you? place the country under martial law?"[54] Byron's anticipations of the extent of the troubles were extreme, but the English army was dispatched and, with little scruple for bloodshed, put the rioters down.

With the demise of alternate ways of relating ends to means in society, the technical intention, according to Ellul, finally ruled supreme. Through decades of successful practice, what emerged was "the technical phenomenon"—a condition in which the consciousness and judgment of all became exclusively oriented to technical solutions. There was, in other words, much more to the triumph of la technique than the effective ordering of new inventions and technical operations. Also involved was the gradual adaptation of *all* human needs, desires, plans, and processes of thought to the technical mode—that is, to the rational, artificial, productive mode of activity. Human consciousness was not absent from the development. It was very much present. But it was now everywhere shaped within extremely narrow technical channels. "The twofold intervention of reason and consciousness in the technical world, which produces the technical phenomenon, can be described as the quest of the one best means in every field. And this 'one best means' is, in fact the technical means. It is the aggregate of these means that produces technical civilization."[55]

Along with writings already discussed, Ellul's work provides an occasion to evaluate the existing speculation on Western technological dynamism. Although such theories differ greatly in substance, they share certain characteristic virtues and defects.

Without doubt, the basic aim is entirely admirable. The questions raised are good ones, even when the answers fall woefully short. They do bring us to reflect upon important tendencies that, evidently, do have their genesis deep in our culture's past. In recent popularizations like Robert Pirsig's *Zen and the Art of Motorcycle Maintenance*, these questions are fruitfully raised for a mass audience ready, in some way at least, to reexamine the modern stance toward things technical. Technological society, after all, has never shown any great commitment to self-reflection, self-criticism, or the study of its own history. Writings of this kind try to fill in what is actually an embarrassing lack in our understanding, the understanding of the nature of the major renovations in civilized life that have taken place in recent generations.

In Ellul's case, for instance, the stress upon the "one best means" can be seen as critique of the radical narrowness so evident in much of technical thought and action. Modern conceptions of practice stipulate that effective performance requires a drastic limiting of the factors perceived or operated upon. One excludes most of the variables present in a situation and deliberately focuses upon a scant few. In identifying this way of seeing and acting, one is not casting aspersions on any particular kind of person, not suggesting, for example, that scientists or technologists are inherently shallow or insensitive. That is not the point at all. As Wylie Sypher has suggested, the technical mode is just as much characteristic in modern arts and letters as in science and technics. Speaking of the progress of art in the nineteenth and twentieth centuries, Sypher observes, "The heavy investment in method suggests that the artist was subject to the same imperatives as the scientist and that the fissure between science and art was not so wide as is alleged, or the kind of fissure one might think, since both were highly specialized executions or procedures."[56] In the technical mode, painting becomes a matter of solving problems of color and form. Rewards are passed out to those who determine how best to paint a white square on a white background. In a technicized profession of philosophy, one uses language analysis to "solve" age-old conceptual problems. Progress is made by those who unravel the standard "puzzles" in epistemology, ethics, or metaphysics. In political science Ellul's "technical phenomenon" occurs in the dominant notion of how the discipline improves: through the continual updating of survey methods, interview techniques, data analysis, concept analysis and a never-ending retooling in the sphere of methodology. The idea that the enterprise had something to do with shared wisdom about politics has gradually become an anachronism.

Fascination with technique could easily be traced through any number of fields of contemporary practice. Success lies in refined performance within carefully circumscribed limits following the proper method. This is just as true of rationalization and systematization in the professions as it is for those who employ techniques of transcendental meditation or make love with *The Joy of Sex* in one hand, their partner in the other.

Much more than the domination-of-nature hypothesis, Ellul's rendering of the "technical phenomenon" helps to reveal much of what is most perplexing in the way modern society goes about its work. Technically educated persons face what they believe to be perilous uncertainty in areas beyond the "scope and method" of their training. They are inclined, therefore, to look no further. As for the broader place of each variety of technical performance within the whole of society, that is quite literally no one's business. It is precisely this well-trained technical narrowness that lies at the foundations of the phenomena of unintentionality and technological drift I pointed to in the previous chapter. The profound depth of this tendency is, I believe, best illustrated by the fact that even those who now acknowledge a problem in man's relations with nature often move from that insight to become unreconstructed technological systems builders on a potentially colossal scale. A noble desire to save the biosphere from destruction is expressed in a frantic whirl of computer modeling, systems dynamics calculations in the Jay Forrester school, and plans for more ambitious global management. Of course, this is the irony of Ellul's thesis—that characteristically almost anything we seize upon to liberate us from our plight simply reinforces the basic quandary.

The strengths of Ellul's vision, however, are matched by equally serious deficiencies. While he pays attention to a wide variety of sociological conditions that contributed to the shaping of modern society, his treatment of them remains relatively superficial. A truly elegant theory of social circumstances in eighteenth-century Europe and America that led to universal adoption of a technical mentality is missing from his book. Instead, the factors he lists all become preconditions of a univocal phenomenon, which can thereafter be treated by single-factor explanation. It is an odd state of affairs that a thinker so wary of mechanism should have produced so mechanical a theory. This reductionist tendency in his thinking, common to philosophers searching for the flaw in Western culture, leads to a badly distorted reading of recent history. Indeed, techniques of communication, propaganda, and police work have played an important part in the rise of modern totalitarian states. Refinement in techniques of warfare definitely has something to do with the frequency and destructiveness of war in the twentieth century. But to explain war, totalitarianism, or any other fact of the times wholly or substantially by reference to the culture's obsession with technique is certainly to overstate the case.

The Search for a "New Ethic"

Characteristic of writings that probe the ultimate philosophical, religious, or historical grounds of technological society is a tendency to gravitate toward extreme hypotheses. Hence, Ellul insists that the "one best means" must be understood in an absolute, rather than a relative sense. To interpret technical improvement as a matter of degree is, he believes, to ignore the very essence of technique. The drift of other important writings in this genre—notably those of Edmund Husserl, Martin Heidegger, and Max Scheler—continue the tendency of representing technology as a totally univocal phenomenon, a monolithic force in modern life.[57] Once the thinker has penetrated the unitary core of this phenomenon, anything else that one might say or do about questions in this area becomes mere child's play. At least so it appears from the perspective of theories of this kind.

Following the implications of the philosophic quest, participants in the discussion have gone in either of two directions. One is the path of thoughtful passivity, which acknowledges the scale of the dilemma and realistically retreats from any thought of activism. Ellul's reflections at the conclusion of his book tell us that: "enclosed within his artificial creation, man finds that there is 'no exit'; that he cannot pierce the shell of technology to find again the ancient milieu to which he was adapted for hundreds of thousands of years."[58] While it is true that part of Ellul's intention is to goad his readers to action, the final remarks give no hint of possible hope.

In a similar way Martin Heidegger's *What Is Called Thinking?* takes the reader on a fascinating journey of ideas which, among other things, promises to answer the question of how man might master technology out of control. Late in the book Heidegger finally reveals to us what can be attained in this search. "We must," he writes, "first of all respond to the nature of technology, and only afterward ask whether and how man might become its master. And that question may turn out to be nonsensical, because the essence of technology stems from the presence of what is present, that is, from the Being of beings—something of which man never is the master, of which he can at best be the servant."[59] A thorough treatment of Heidegger's theory of technology and its basis in existential phenomenology is something I must forego here. Nevertheless, one can appreciate from the above passage that Heidegger's speculations tend to make any question of the sort "What next?" largely irrelevant.

A second route that philosophers have traveled is to propose the adoption of a "new ethic" for technical and scientific practice in our culture. This solution seems particularly attractive to those who have decided that the root malady lies in Western man's relation to nature. Lynn White finds the basis for a revised moral stance in the teachings of Saint Francis of Assissi.[60] An alternative Christian ethic would, he maintains, recognize the equality of all creatures and discard the notion of human monarchy over nature. Following a neo-Marxian approach to much the same problem, William Leiss takes a bold step that Horkheimer and Adorno, perhaps wisely, avoided:

The idea of the mastery of nature must be reinterpreted in such a way that its principal focus is *ethical or moral development* rather than scientific and technological innovation. In this perspective progress in the mastery of nature will be at the same time progress in the liberation of nature. The latter, like the former, is a rational idea, a concept, an achievement of human thought; therefore the reversal or transformation which is intended in the transition from mastery to liberation concerns the gradual self-understanding and self-disciplining of *human* nature.[61]

These are noble sentiments. But one is left wondering what kind of fresh resolve such writings actually advocate. In White's case, for example, it is difficult to believe that following some astounding reevaluation of the culture's latent religious foundations, men in Western societies would take up Saint Francis's beneficent countenance. Yet that is exactly what White appears to suggest. After reading Leiss's statement of advocacy, one is still perplexed as to what "progress in the liberation of nature" might mean. That human beings can be liberated makes sense. Perhaps that is why Leiss ties his point so closely to "human nature." But those who have worried about man's "domination," "despotism," or "mastery" over nature have had much more than human nature in mind. At stake is nothing less than the capacity of modern science to bring all of material reality under intellectual and practical command. The play of conceptual opposites in Leiss's formulation—domination versus liberation—leads to confusion when applied to man's heavy-handed treatment of the environment. One may wish to counsel something. But "liberation" does little to identify it.

The expectation that a rapid shift of attitudes could follow a philosophical critique is also badly misplaced. What would such a change look like? I imagine posted signs that read:

Attention friends and neighbors. Today begins a fundamental change in our view of ourselves and the biosphere. Henceforth, instead of acting as predators upon the land, the plants, and animals, we will take a more enlightened, self-disciplined, moderate stance. The very next time you walk up to a tree, flower, cow, bluejay, or babbling brook, assume a peaceful feeling in your heart of hearts. (And so forth.)

The prudence of philosophy has always included an understanding that prescriptions of too great a specificity tend to be absurd. That is one reason why Marx remained silent on the design of communist society, why Plato left much unsaid about the path to the Form of the Good. Even the deepest insights about the world and man's place in it may, nevertheless, leave one perplexed as to what action is appropriate. To have completed even the most effective critique of Western metaphysics is no guarantee that one will know what, if anything, to do next. The great mistake of attempts to propose a "new ethic," it seems to me, is that they proceed as if the enterprise were something like philosophical engineering—a special kind of problem solving. But at the level at which one seeks to elucidate first principles about man, nature, and being, the world is not a problem set. It is, if anything, a question set.

In summary, explorations of the origins of excess and uncontrollability in modern technological practice suffer two important shortcomings. First is an eagerness to advance unitary, totalistic hypotheses with unwarranted conviction in an area marked by a multiplicity of complicated questions and a high degree of uncertainty. The theories pretend to have knowledge that is not so easily available and to which, possibly, we are not even entitled. Second is a penchant for suggesting vast revolutions in consciousness where good sense and moderation might do. An overestimation of what the critique reveals is matched by odd conceptions of what action is appropriate.

Surely there must be a more fruitful way of going about this, a more helpful approach to the questions that matter. Rather than storm the metaphysical foundations of civilized existence hounding an illusory "new ethic," it may be more to the point to reexamine a number of traits most closely linked to the development of Western technology—the spirit of scientific inquiry; the vital interest in invention, tinkering, fixing, and problem solving; the risk-taking, gambling urge of innovators and entrepreneurs (often mistaken for a mere lust for profit); the universal desire to rid the world of pain and trouble—and then ask when and how such impulses get out of hand. No evidence exists of any culture that has gotten away without some attempt to understand, alter, and exploit nature. No evidence can be found of any human society that has not employed tools and techniques. The interesting question is why the modern West has proceeded along these paths with virtually no sense of limits.

If I am right, what is needed are inquiries that stand somewhere between the ultimate "Being of beings" and the latest squabble on this or that social gadget. One must seek simultaneously to avoid depths without direction and details without meaning. It is with this understanding that I now turn to a subject that has traditionally found its work in the middle ground between ideas sublime and affairs quotidian—the subject of political theory.

Chapter 4
Technocracy

Many of the problems we have been considering are material for politics. Technological change is now widely recognized as political insofar as its effects are ubiquitous, touch everyone in society, and can, therefore, be understood as "public" in a distinctly modern sense. Understood in this fashion, political questions concerning technology abound at every turn. One can study the effects of satellite communications on international affairs or of computers upon the style and content of political campaigns. One can examine the interest-group struggles that stand behind the approval of an antiballistic missile system or the veto of a supersonic transport program. Increasingly, social scientists involved in policy studies find themselves confronting such issues, for the domain of politics and that of technics overlap.

But without denying the importance of such research, I want to follow a different approach to the intersection of the two spheres. The image of autonomous technology warns us to avoid any premature narrowing of our conceptions of either the political or the technological. To place the issue in terms of a single case study or the behavior of a particular agency, community, congressional committee, or set of interest groups is to ignore a range of problems which beckon us. Our next step, therefore, will be to examine two alternate ways of seeing technology as a general dilemma in political life. The first approach, the more conventional, considers the question of technocracy—rule by scientific and technical elites—and its challenge to traditional thought and practice.

Classical Technocracy

It is unfortunate that the first modern vision of a technocratic society, Francis Bacon's scientific utopia, *New Atlantis,* was left unfinished at the author's death. Written in the period following Bacon's ignominous fall from office, the book is tentative or ambiguous on most of the questions that now seem most important, particularly those concerning the relationship between scientific and political power. The forty or so pages of the book describe the visit of a crew of English sailors to the island kingdom of Bensalem. Most notable and esteemed of Bensalem's institutions is a magnificent scientific academy, "Salomon's House" or "The College of the Six Days Works," which the visitors are told "is the very eye of the kingdom."[1] Clearly patterned on the design of collective scientific research and technological application outlined in the *Novum Organum*, the purposes of Salomon's House are exactly those one would expect from a Baconian scheme. "The End of Our Foundation is the knowledge of Causes, the secret motions of things; and the enlarging of the bounds of Human Empire, to the effecting of all things possible."[2] The man of scientific ambition, who we noticed earlier, has now been set to work building what Bacon had described as a nonpolitical "empire." Throughout the kingdom, stations for scientific research and development have been established, manned by highly trained crews. Every conceivable aspect of nature that might be probed or modified is given the most rigorous attention. This work has continued since its founding nineteen hundred years earlier by a man with a "*large heart*, inscrutable for good," King Solamona, who "was wholly bent to make his kingdom and people happy."[3] The fruits of this work are now manifest: vast knowledge of the earth, air, water, animals, fish, and vegetation along with the useful development of great machines, foods, and marvelous devices of every description.

But there is something very peculiar about the land of Bensalem and much in *New Atlantis* that leaves us wondering. A strongly enforced veil of secrecy shrouds most of the life of the island. There are certain things about which the "strangers" simply may not learn. All that one knows of the country after reading the sailor's account is that it is a Christian nation and that it contains the scientific academy. There are occasional references to "the state," to a king and governors of cities, as well as to other minor state officials. But the character of this state and its relation to the scientific foundation is not spelled out, although the two are apparently distinct entities. The Father of the House, introduced to the visitors, remarks that there are certain findings that "we think fit to keep secret: though some of those we do reveal to the state, and some not."[4] The visitors repeatedly note that Bensalem is a happy place, but what little glimpse of the people of the kingdom we receive is one of a regimented, rigidly obedient lot obsessed with religious piety, secretive in all of their dealings, and fascinated by the possibility that somewhere in the world there are sinful, unchaste goings-on. At the ritual of the "Feast of the Family," the people recite the slogan: "*Happy are the people of Bensalem.*"[5] But why they should say this is never evident.

Unlike other utopian writings, there is almost nothing in *New Atlantis* to back up the claim that the world described is happy or well governed. Bacon's obvious model for the work, Thomas More's *Utopia,* does spell out the nature of the political institutions of an imaginary land and makes clear exactly why its people are contented. Remarkably, Bacon does none of this. He refuses, at least in this fragment, to provide the endless details with which most other utopian writings are filled. As a result, the land takes on an air of mystery and even foreboding. Is Bensalem a good state or a bad one? We do not know. But the now-standard interpretation of *New Atlantis* as a perfectly ordered scientific utopia is certainly open to question.

What Bacon would have done with the book if he had finished is an interesting question for speculation. One theme he may have wanted to develop is that of the political world as one of corruption and ineptitude and the realm of science, in contrast, as one of purity, intelligence, and competence. Evidence of corruption in Bensalem appears when the visitors offer gifts to the people. In each case the people refuse the gifts and are shocked at the attempt to give them. "What? twice paid!" one of their guides responds when the visitors hand him some money. For "they call an officer that taketh rewards, *twice paid.*"[6] Why are the people distressed about such offerings? Is there a history of scandal in the politics of the land? Much of this aspect of the book may be a satire on the bribery charges that brought Bacon's own downfall. In the words of the Bensalemians, Bacon was himself twice paid in the custom of his time and later suffered for it. The references to possible bribery in *New Atlantis* could be dismissed lightly were it not for the fact that the fragment ends with the Father of Salomon's House giving a sum of money to the visitors. "And so he left me; having assigned a value of about two thousand ducats, for a bounty to me and my fellows. For they give great largesses where they come upon all occasions."[7] Why do the Fathers of the House give great sums of money when they come and go? Does this mean that when men of high virtue in the realm of science and technics have dealings with men in the mundane or political world, they must resort to bribery? Or are they somehow beyond all of that and able to make their generous gifts for different reasons?

It is conceivable that if Bacon had finished his utopia, the line between scientific technology and politics would have been more clearly drawn, probably at the expense of the latter. Written three decades before the founding of the Royal Society, *New Atlantis* may have been a prophecy that scientific and technical institutions were uniquely suited to govern in the age of "The Great Instauration." Politics of the traditional sort could only be a drag on the development. In this regard the book does introduce two major themes that run throughout modern technocratic writing: power and authority placed on a new foundation.

If there are other sources of political power equal to that of Salomon's House, we are not told of them. The works of the foundation are the guiding influence in all the affairs of Bensalem. The political sphere has become a matter of administration on the one hand and of hollow ritual and symbol on the other. True governance in the land is contained by the knowledge, organization, and performance of the scientists and technicians, who apparently are given free rein.

Salomon's House and its ruling elite have also gained a predominance of authority. When one of the Fathers of the House visits a city he is feted with ceremony due an emperor. "Behind his chariot went all the officers and principals of the Companies of the City. He sat alone, upon cushions of a kind of excellent plush, blue; and under his foot curious carpets of silk of divers colours, like the Persian, but far finer. He held up his bare hand as he went, as blessing the people, but in silence."[8] The inhabitants of the city line up along the streets and at the windows "as if they had been placed" to offer homage to the great man. The

Father, with "an aspect as if he pitied men," is the embodiment of aristotechnocratic rule.[9]

Conceptions of power and authority in technocratic writings have remained virtually unchanged since Bacon. Power is ultimately the power of nature itself, released by the inquiries of science and made available by the inventive, organizing capacity of technics.[10] All other sources of political power—wealth, public support, personal charisma, social standing, organized interest—are weak by comparison. They are anachronisms in a technological age and will ultimately decline as scientific technology and the people who most directly control its forces become more important to the workings of society. To say this is only to recognize the drift of history and to be realistic. Elaborate political facades of various kinds may still surround the exercise of technocratic power, but beneath the surface, the true situation will always be evident.

Authority, like power, is in this point of view the product of knowledge and extraordinary performance. If those persons valuable or indispensable to the policy are those entitled to govern, then a society based on sophisticated technologies will tend to legitimate its scientists and technicians as rulers. Their expertise and accomplishment will naturally gain the esteem of the other members of society. Other sources of authority from earlier times and earlier understandings of the common good—tradition, religion, natural law, contract—must inevitably yield to this new mode of legitimation. In the search for an appropriate comparison, some writers have employed the metaphor of the priesthood to try to capture the special authority of the new men of knowledge.[11]

In most actual statements, these ideas are not presented in anything like a pure form. They are modified by numerous other political concerns and circumstances, some of which we will examine later on. But the technocratic strand of thinking, insofar as it is an influence on modern political theory and whether it speaks in favor, opposition, or neutrality, always rests on premises of the kind we have just seen.

These are, of course, basically unpolitical ideas. They stand to the political tradition of the West in much the same way that General Motors stands to a blacksmith. If the possibility they express is accurate, one could only look forward to a spectacular supersession of politics (whatever its definition). The power described here is the cancellation of all other varieties of power and the cancellation of the historical debate about how power exists and how it works. The authority rests on a human population dwarfed and submissive before forces it cannot understand or influence but entirely content with the services offered.[12] Even the most pessimistic of traditional thinking on man and society has granted more to the public sphere than this.

The occasion for technocratic thought has usually come in one of two ways: as utopian speculation or as commentary on the devolution of existing political systems. In the first instance, technocracies appear in designs for a world of the future in which the affairs of men have been perfected through a rational, harmonious plan. In the second, such notions are the conclusions of men who have read the currents of history and find that, for better or worse, society will inevitably be governed by scientists and technicians of one kind or another. Here is another reflection of the modern ambivalence about technology. A technicized regime deliberately built according to an ideal plan is the penultimate dream of all science fiction and science fantasy. But technocracy which arrives through no settled plan but as the result of the enervation of politics and the rise of a technical elite is one of the primary vexations of modern political thought. At one time or another technocratic rule has been announced as the logical successor for every sort of political and economic system: feudalism, democracy, capitalism, socialism.

In many instances utopian and historical speculations have been combined. The demise of a political system is seen as an opportunity for the building of a technological society ruled by a technically competent aristocracy. This was the outlook of Henri Comte de Saint-Simon at a time when the ancien regime was being dismantled and a new system constructed in its place. Saint-Simon's criticism of the French Revolution was that its efforts were overly political and did not take into account the realities of the new mode of social organization taking shape at the same time. "The men who brought about the Revolution," he observed, "the men who directed it, and the men who, since 1789 and up to the present day, have guided the nation, have committed a great political mistake. They have all sought to improve the governmental machine, whereas they should have subordinated it and put administration in the first place."[13] True progress was located in the development of the new instruments of technology and techniques of governmental administration. This required, Saint-Simon argued, a system of expert management by industrialists, scientists, and technicians.

The precise form of the proposed government was one that now seems very traditional indeed. Saint-Simon placed the members of his technical elite in a parliament with three houses: the Chamber of Inventions, Chambers of Review, and Chamber of Deputies. The Chamber of Inventions, composed of two hundred engineers and a scattering of poets, painters, architects, and musicians, would decide the basic plan for all of France. The Chamber of Review, made up of mathematicians and pure scientists, would judge programs devised by the Chamber of Inventions and serve as a control over its policies. Completing the arrangement of checks and balances, the Chamber of Deputies, composed of practicing industrialists, would serve as an executive body to implement the plan.[14] Notably absent from Saint-Simon's scheme is any trace of equality or electoral democracy. The members of the parliament were to be chosen according to professional competence alone and not elected by the populace at large. The ascendance of scientific and industrial classes could take place only at the expense of a total neutralization of the political role of the majority of men and women, benighted souls, who did not possess higher knowledge and skill. "A scientist, my friends, is a man who predicts," Saint-Simon announces. "It is because science has the means of prediction that it is useful, and makes scientists superior to all other men."[15]

In the decades since Saint-Simon offered his plan, the conception of who would have membership in the technical elite has been repeatedly revised. With the coming of each new technique or apparatus there has usually been someone eager to write a manifesto or utopian novel telling how an aristocracy linked to the development will revolutionize all social and political practice. A noted pioneer in such speculative writing was H. G. Wells. In his "scientific romances" Wells often began by constructing an imaginary society dependent upon a particular kind of advanced machinery and governed by scientists, technicians, or managers of good or evil stripe.[16] With such conditions as a basis, his stories then unfold.

Wells's *When the Sleeper Wakes* takes place in an advanced industrial society in the year 2100 managed by a small class of aviators and engineers.[17] The workers, still miserable and exploited, rise up against the aeronauts and their ruthless leader Ostrog, and in a scene that now seems amusingly old-fashioned, battle the "aeroplanes" to the death. The plot is a cross between science fiction and Marx. Wells, the first professional futurologist struggling with the social implications of advanced technology, experimented with numerous such schemes in his fiction. Without exception, his works tend toward the same political conclusion: that the conditions of the modern age would require rule by a relatively small, cohesive group of highly trained technicians. While he feared the evil this kind of government might bring to the world, he eventually came to believe that the inevitable development would also be a blessing. The final reconciliation came in his book *A Modern Utopia.*[18] Here we find a society in which the ruling class is entirely benevolent and is no longer defined by its attachment to a particular kind of machinery. The "samurai" are a well-rounded, intelligent, self-selected, ascetic class of managers educated in all of the skills required to run a scientific society. The condition produced is a kind of social euphoria, a pallid perfection. Wells's utopia is self-consciously modeled on Plato's *Republic* and Thomas More's *Utopia.* The samurai are actually Platonic guardians who have seen the Forms, not of Justice or the Good but the forms of efficiency and technical order.

In addition to the stream of fictional utopias, there have been a number of straightforward works leading the cheers for the extraordinarily gifted children produced in the womb of modern technique. F. W. Taylor's *Principles of Scientific Management* is, among other things, an argument in justification of a new class of managers, which Taylor believed indispensable to all twentieth-century industry.[19] A more recent panegyric, Robert Boguslaw's *The New Utopians,* serves a similar role for the systems analysts, although under the guise of warning the public about the possible dangers of such people. In technocratic writings, a stern warning is often a rhetorical style to mask wholehearted endorsement, for indications of new power carry much weight, while halfhearted bemoaning of moral dilemmas falls flat.[20]

At least as interesting as the matter of who belongs to the technical elite is the question of who does *not* belong. The literature contains occasional purges within the pantheon, with some esteemed members finding themselves cast out. Saint-Simon's conclusion that "the industrial class is bound to continue its progress and finally to dominate the whole community"[21] was not one that Thorstein Veblen could endorse a few decades later. Veblen held that the progress of the modern system of industrial production had reached a point at which it no longer contained any crucial function for the men formerly most important in the building of that system—the "captains of industry," businessmen and financiers. Because such men were addicted to "a strict and unremitting valuation of all things in terms of price and profit,"[22] they were inherently unable "to appreciate those technological facts and values that can be formulated in terms of tangible mechanical performance."[23] The interests of the old industrial governing class caused it to run the industrial system at a low level of output and efficiency. Society as a whole paid the price for this selfish manipulation, for it was deprived of the full product of industrial technology. Veblen felt no qualms about labeling the activity of the business elite deliberate "sabotage." "The cares of business have required an increasingly undivided attention on the part of business men, and in ever increasing measure their day's work has come to center about a running adjustment of sabotage of production."[24] In complicity with this disgraceful practice were the politicians, working on both a national and international scale to direct the riches of the productive network to the "vested interests." "So it happens," he explained, "that the industrial system is deliberately handicapped with dissension, misdirection, and unemployment of material resources, equipment, and man power, at every turn where the statesmen or the captains of finance can touch its mechanism."[25]

By any standard of intelligence or justice, Veblen believed, the true rulers of modern industrial society ought to be the technicians directly responsible for the efficient running of the machine. The engineers were actually in control anyway. It was high time to recognize their importance and to free them from the sabotage from above, which prevented their full efficiency.

In more respects than one the industrial system of today is notably different from anything that has gone before. It is eminently a system, self-balanced and comprehensive; and it is a system of interlocking mechanical processes, rather than of skillful manipulation. It is an organization of mechanical powers and material resources, rather than of skilled craftsmen and tools; although the skilled workmen and tools are also an indispensable part of its comprehensive mechanism. It is of an impersonal nature, after the fashion of the material sciences, on which it constantly draws. It runs to "quantity production" of specialized and standardized goods and services. For all these reasons it lends itself to systematic control under the direction of industrial experts, skilled technologists, who may be called "production engineers," for want of a better term.[26]

Veblen lamented the fact that the class of engineers had not yet achieved full consciousness of its role and potential power. They were not organized to seize control from the captains of industry and, therefore, "under existing circumstances there need be no fear, and no hope, of an effectual revolutionary overturn in America, such as would unsettle the established order and unseat the Vested Interests."[27] At the same time, Veblen thought it impossible that the practice of sabotage could continue forever. Eventually the close interconnection and interdependence of the evolving industrial system would make such jamming and tampering the source of a crisis that would "bring the whole to a fatal collapse."[28] He hoped that before that day arrived, the noble engineers would come together around a common plan of action, possibly a general strike, and assume the position of power and authority they deserved.

But whatever changes occur in the ideal membership of the technical elite, there is always one group excluded from the list—the great mass of men in society. Most persons, it is asserted, simply lack the knowledge or credentials to participate in the government of a technological society. The technocrats rule because no one else is capable. In the last pages of *The Decline of the West*, Oswald Spengler emphasizes this side of the technocratic argument, calling attention to implications that Veblen had no doubt noticed but played down. "The peasant, the hand-worker, even the merchant," he writes, "appear suddenly as inessential in comparison with the *three great figures that the Machine has bred and trained-up in the cause of its development: the entrepreneur, the engineer, and the factory-worker.*"[29] But two of these "figures" have already been obscured by "a mighty tree that casts its shadow over all the other vocations—namely, the economy of the machine industry. It forces the entrepreneur not less than the workman to obedience. Both become slaves, and not masters of the machine, that now for the first time develops its devilish and occult power."[30]

But there is still one man, Spengler contends, with sufficient knowledge and skill to handle this new power. He alone is able to rise above the common ignorance and the "conflict of politics" to become ruler of a mechanical civilization. "Not merely the importance, but the very existence of industry depends upon the existence of the hundred thousand talented, rigorously schooled brains that command the technique and develop it onward and onward. The quiet engineer it is who is the machine's master and destiny."[31] In a later work, *Man and Technics*, Spengler is even more definite about the authoritarian implications of the gap between the technically proficient engineers and benighted masses. This is, he explains, merely the newest instance of an age-old relationship between rulers and the ruled. "As in every process there is a technique of direction and a technique of execution, so, equally self-evidently, there are *men whose nature is to command and men whose nature is to obey, subjects and objects of the political or economic process in question.*"[32] Technocratic rule was for Spengler the last gasp of a culture entering the advanced stages of decay. His call to the heroic engineer was actually a dirge sung very loudly.[33] It was to this song that Albert Speer responded, bringing with him the Spenglerian-technocratic elements of nazism.

Technocracy and Liberalism

If Bacon, Saint-Simon, Veblen, Spengler, and others are correct, rule by technically trained experts is the only kind of government appropriate to a social system based on advanced science and technology. Technocracy in some form or other is the inevitable, rightful heir to all varieties of politics. In the following paragraphs I want to examine the implications of this view for liberalism. A later chapter will consider somewhat similar problems in a Marxist context.

A matter of theory, technocracy's challenge to liberalism is simple and direct. Its premises are totally incompatible with a central notion that justifies the practice of liberal politics: the idea of responsible, responsive, representative government. In the technocratic understanding, the real activity of governing can have no place for participation by the masses of men. All of the crucial decisions to be made, plans to be formulated, and actions to be taken are simply beyond their comprehension. Confusion and disorder would result if a democratic populace had a direct voice in determining the course the system would follow. Science and technics, in their own workings and in their utility for the polity, are not democratic, dealing as they do with truth on the one hand and optimal technical solutions on the other.

This still leaves open the possibility that the populace could voice its desire for the goals and kind of distribution that a system run by experts would obtain. Voting would reflect the wishes of the people, which could then be enacted in the technically best way. But at least in the formulations we have seen so far, this is not at all the way that theorists of technocracy have seen the matter. They consider the accomplishments of the new technology so obvious and marvelous that they expect the public to receive them willingly and without asking questions or making demands. Why preprogram cornucopia? The technocrats, being human themselves, understand man's basic needs. They do not need to be reminded of them. In a world being transformed and made better through new devices and techniques, the voice of the public can only be a kind of ignorant carping.

For liberal politics this is an unacceptable position, guaranteeing neither good government nor safety. What is required is some means of representation to make certain that the wishes of the public are followed and the power of the experts limited. Where the populace as a whole may not be capable of dealing with the complexities of new technical systems and the kinds of knowledge and decision they require, their representatives will be able to master this field and maintain liberal democratic government. The technocratic response would be that this notion of representation is a pipe dream dreamed by those whose pipes are already clogged. Delegates of popular will must inevitably meet the same fate as the populace itself—functional irrelevance. The elected spokesmen may still be present and voting, pretending to legislate the goals of the system and acting as if they exercised a final check on what is done and how. They are entitled to this kind of entertainment. But its efficacy is, at best, very limited, for the knowledge, power, and authority of the traditional politician are not actually necessary to make the system run. Political office becomes a collector's item, a glorious antique.

It is time to ask about the reality these ideas describe. How well do actual cases match the speculation?

There is a standard means of inquiry available to us. Taking the works cited above as examples, technocratic writing can be shown to be a subcategory of an important branch of modern political thought and social science: the theory of elites. Investigations begun by Vilfredo Pareto, Gaetano Mosca, and Robert Michels and continued more recently by C. Wright Mills, Floyd Hunter, G. William Domhoff, and innumerable political scientists have sought to identify the conditions under which elites do or do not exercise political influence, the relationships elites have to other social strata, and the conflicts among competing elites.[34] I shall not enter into the details of this now-complicated discussion but merely observe that technocratic writings point to one possible group that fits Pareto's original definition of a *governing elite*, "individuals who directly or indirectly play some considerable part in government."[35] This is exactly what Bacon, Wells, Veblen, Burnham, and others have pointed to and precisely what Dwight D. Eisenhower meant when he spoke of the "danger that public policy could itself become the captive of a scientific technological elite."[36]

In their research, political scientists examine a number of signs to determine whether an elite is present and active in a given situation. Members of the elite have (1) similar social backgrounds, (2) a common ideology, (3) a desire for generalized power, (4) the ability to communicate with other members and to act in a unified manner, and (5) direct access to positions of command in society. An elite, in other words, is usually defined as a cohesive social group which intentionally seeks political power.

Without laboring the point, social scientists have been unable to find a technocratic elite to match anything like this description in any existing liberal democratic society. Looking at the persons who might have been material for a Baconian or Veblenian ruling class—scientists, technicians, technologists, industrial managers, bureaucrats, and the like—one does find varying degrees of access to and use of political power.[37] Holders of expert knowledge are found actively engaged in many kinds of planning, advising, administrating, decision making, and lobbying.[38] And it is commonly noted that persons of this sort do have an influence in political affairs disproportionate to their relatively small number. Nevertheless, in terms of the categories used to judge political elites, men of scientific and technical expertise do not fit the model. They do not share a common social position, uniform ideology, or any sort of ruling class consciousness. They neither have nor perceive the need for solidarity and do not behave as a cohesive, power-seeking group. The idea that they might want generalized social or political power seems entirely foreign to them. On the contrary, the concerns of such people are tied to their specific professions. They do organize and pursue their interests within the professional setting, and often this means that they involve themselves in the political sphere, but their activities here are limited and usually highly specific. Far from wanting to govern, they provide services, command a price, and leave the rest alone.[39]

All of this has been studied in considerable depth by political scientists. In the United States the spotlight fell primarily upon the community of scientists and its post–World War II influence. In France the discussion centered directly upon the "technocrats," all technically trained persons in or near positions of political power. Scholars in both countries paid attention to the possibility of indirect as well as direct influence by scientific and technical elites. The standard conclusion was (and still is) that none of the kinds of people identified as possible candidates for a technocratic ruling stratum actually form a cohesive, unified elite to interfere with the workings of liberal democratic government. Victor Ferkiss writes, "The scientific estate is now one among many in a pluralistic struggle for individual and group power, and far from the strongest."[40] Jean Meynaud concludes, "Technocracy has not managed to gain a completely preponderant control of government action in any contemporary regime, supposing that this is in fact the true wish of technicians."[41]

If one expects to find technocracy in the sense described in the early writings in this lineage, one will be disappointed. In the American experience nothing symbolizes its nonexistence better than the tragic-comic history of Howard Scott's Technocracy Inc. Largely founded on the ideas of Veblen in *Engineers and the Price System,* the technocratic movement aimed at reorganizing the whole American economy according to a comprehensive master plan implemented by the best technicians, a plan that would increase productivity to a new high and guarantee abundance for everyone. But the ideas and grand symbol of the movement—the red and silver, yin and yang "Monad"—never caught on. Its caravans of uniform gray automobiles carrying hundreds of gray-suited men made coast-to-coast treks to publicize the goals of Technocracy Inc. But other than some enthusiastic support in such political centers as Pismo Beach, California, in the 1930s and 1940s, the whole thing was ignored by the public and, much worse, by the noble technicians themselves. In its best moments Technocracy Inc. was an organization of crackpots; in its worst, an inept swindle.[42]

But the failure of technocracy in one definition—the definition suggested by the theory of elites—does not mean that the power and position of technically trained persons in political life ceases to be a problem. That there is apparently little solidarity or common purpose among such persons does not in itself speak to the issues raised above about participation, representation, or limited government. It merely denies one possible way that technology and political power might be connected. The elite conception of technocracy is, it seems to me, a good example of a case in which "a *picture* held us captive."[43] The idea here is that of a cohesive group based on the knowledge it holds *rising* to power and authority. In science fiction and political theory both, there is a tendency to dramatize the upward thrust, hence the titles "new brahmins," "new mandarins," "new priesthood," and "new utopians." And if one sees society in terms of strata or class levels, the strongest being on "top" and the weakest at the "bottom," then one begins to expect that those who hold new social power will move "upward" and like mountain climbers at the top of Everest be somehow visible up there.

But this conception is much too simple. Contemporary students of politics, including those interested in the relationships between science, technology, and political power, have had to deal with a much more intricate situation. They have now recognized that there is a multiplicity of kinds of persons who hold power and that accompanying this is a relatively wide diffusion of power within the political system. In mentioning this, I do not wish to affirm the branch of normative political philosophy known as pluralism. The pluralist looks at social diversity and the balancing of interest group pressures and concludes that he has discovered democracy. In the multiplicity of power holders and diffusion of power he sees signs that the republic is in fine fettle. But as critics of pluralism have argued, this is a drastically crippled notion of democratic politics, leaving as it does questions of equality and participation out of the picture.

No, in citing this condition of multiplicity and diffusion I am merely noticing something that most social scientists, even the proponents of the "power elite" thesis, now admit: that one who looks for power in society does not find a single, homogeneous group at the "top."[44]

Awareness of this state of affairs has influenced a body of mid-twentieth-century writing—perhaps one could call it revisionist technocratic theory—which attempts to redefine the relationships between technical expertise and political power and authority. I want to consider briefly two of the best thought out of these new statements: Don K. Price's discussion of the scientific estate and John Kenneth Galbraith's description of the technostructure. We shall see that while the problem is not as clear-cut as it was formerly conceived, technocracy is still a source of considerable vexation for liberal politics.

The Scientific Estate responds to what Don Price perceives to be a profound dilemma for the American constitutional system. The rise of science and scientific technology in the nineteenth and twentieth centuries has, he believes, altered the basic rules and procedures of U.S. government established at the time of the founding. In particular, there have been three developments that make "our traditional reactions—our automatic political reflexes—unreliable in dealing with our present problems."[45] The scientific revolution has (1) moved the public and private sectors closer together, (2) brought a new order of complexity to the administration of public affairs, and (3) upset our system of checks and balances in government. Price's concern with this state of affairs moves in two directions. Like Kant, who wished to save morality from the rule of science and science from the rule of morality, Price thinks it imperative that politics and science be maintained as distinctly separate kinds of activity. The new threat to politics, he suggests, is that scientists might take their admittedly powerful ways of knowing and acting and use them to establish supremacy over traditional political institutions. The danger to science, on the other hand, is that politics will too strongly assert its governing function and attempt to interfere in the quest for scientific truth. Price holds that, as a matter of fact, a new sort of constitutional system is evolving to handle the relationship of established legal authority and the scientific professions. The principles of the new American constitution are set forth for the first time in his book.

The genesis of the system takes place in a national environment in which there has been a *fusion* of economic and political power. This is a conclusion that Price shares with many students of the post–World War II relationship of business and government. In those sectors of the economy involving large, sophisticated scientific technologies, the boundary between private industry and government has become increasingly indistinct, in some cases virtually nonexistent. Price portrays this matter in as benign a light as possible, for he is certain that both government and highly technicized business are inherently responsible, benevolent actors. "Today, our national policy assumes that a great deal of our new enterprise is likely to follow from technological developments financed by the government and directed in response to government policy; and many of our most dynamic industries are largely or entirely dependent on doing business with the government through a subordinate relationship that has little resemblance to the traditional market."[46] This is the realm of "big" science and technology, heavily financed by federal tax money, characterized by multimillion dollar contracts, best illustrated, as Price points out, by the development of weapons systems for the nuclear-aerospace age. "The very uncertainty of the research and development process requires the government and business to work out a joint arrangement for the planning and conduct of their programs; the relationship is more like the administrative relationship between an industrial corporation and its subsidiary than the traditional relationship of buyer and seller in a free market."[47] Price holds this to be an indelible fact of modern life with which it would be petty to quibble. "In the modern industrial world," he says, "there is no way to keep government and business from being dependent on each other."[48] By helping to bring about the "blurring of boundaries," science has actually "changed the nature of property."[49] Price concludes: "In short, we are now obliged to think about the political functions of various types of people not on the basis of the property they own but of what they know, and of the professional skills they command."[50] Here we find a revival of one of the premises of traditional technocratic theory: the supremacy of knowledge over property and the eclipse of traditional capitalism on this account. Along with Robert Heilbroner, Daniel Bell, and others, Price has observed and seeks to justify this state of affairs. Such findings are invariably offered as new and unprecedented insights, as if the renegade, discredited Veblen had never set pen to paper.[51]

The *fusion* of economic and political power is accompanied by an equally significant development which Price labels a *diffusion* of sovereignty. No longer does sovereign authority rest solely with the legislative and executive branches of government or with the political parties that control them. Instead, the "process of responsible policy making . . . is a process of interaction among the scientists, professional leaders, administrators and politicians; ultimate authority is with the politicians but the initiative is quite likely to rest with others, including scientists in or out of government."[52] This interaction of different kinds of policy makers is what Price means when he speaks of the new order of complexity which has overtaken public affairs. The authority to make policy and to act is now much more spread out than previously. It involves many more occupational and professional groups with many more kinds of relationships than are spelled out in the written Constitution. This new web of governmental connections and associations took shape in response to specific historical circumstances during the last several decades, the most important of which are a function of the power and utility of modern scientific technology.

The task of Price's theory is to delineate the general principles upon which the "unwritten constitution" brought to America by the scientific revolution now operates. How does the new system of power and authority work? Is it as it ought to be? In specific, Price wants to know whether it retains the quality of good, responsible, representative democracy.

Price continues by spelling out the fundamental divisions within the new order of diffuse sovereignty. There are, he maintains, four functional groups within the system with different spheres of interest, different kinds of expertise, and different sorts of legitimacy. Employing a notion from constitutional theory of the Middle Ages rather than that of the American experience, he suggests that these groups are best thought of as "estates." There is the political estate consisting of elected political leaders, endowed with traditional legal authority; this is the estate most directly responsible to the public. Then there is the administrative estate comprising the managers and administrators in both government and private corporations. A member of this group "is obliged to deal with all aspects of the concrete problems that his organization faces" and "must be prepared to use a wide variety of professional expertise and scholarly disciplines, as he helps his political superiors (or the directors of a business corporation) attain their general purposes."[53] Next is the professional estate, the estate most properly called technological in the sense that I have been using here. This estate is composed of the organized professions—Price cites engineering and medicine as examples—and deals with the practical application of scientific knowledge. The responsibility of a professional "to an individual client or to a corporate or governmental employer, is to serve within a definite and limited field; within that field, the professional has an obligation to standards of ethics and competence that his profession, not his employer, dictates."[54] Finally, there is a scientific estate, composed of scientists in the university, the corporations, and government, doing pure research on the frontiers of knowledge. Their responsiblitiy is to the truth as pursued by the best methods and judged by the highest standards of the scientific disciplines.

The four estates are the approximate equivalent of the three branches of government of yore. Price notes that the lines that divide the estates are by no means hard and fast. "Every person, in his actual work, is concerned to some extent with all four functions."[55] Politicians sometimes have an interest in and a knowledge of science, for example. "The four broad functions in government and public affairs—the scientific, the professional, the administrative, and the political—are by no means sharply distinguished from one another even in theory, but fall along a gradation or spectrum within our political system. At one end of the spectrum, pure science is concerned with knowledge and truth; at the other end, pure politics is concerned with power and action."[56]

It is from this idea of a spectrum from truth to power that Price derives an ideal of good government appropriate to the times. He announces the following principle: "(1) The closer the estate is to the end of the spectrum that is concerned with truth, the more it is entitled to freedom and self-government; and (2) the closer it gets to the exercise of power, the less it is permitted to organize itself as a corporate entity, and the more it is required to submit to the test of political responsibility, in the sense of submitting to the ultimate decision of the electorate."[57] This is the preamble of Price's unwritten constitution. The occasion is not that of a new founding, for the arrangement that this principle helps us to evaluate has grown in bits and pieces over many years. It did not originate in anyone's design. Nevertheless, Price is the Madison of the new system. He wishes to describe the workings of its mechanism and determine its worth. Thus, the question becomes, How well does the principle of the spectrum from truth to power actually turn out in practice? Does it account for what actually takes place?

As Bernard Crick pointed out long ago, the distance between the ideal and the actual in political scientists' views of American politics is never very great.[58] In Price's case there is no distance whatsoever. The spectrum principle describes the existing state of affairs exactly! The scientists are provided with the freedom and self-government necessary for their work, but they do not exercise any direct political authority or power. The politicians are still held responsible by the electorate; they do not have complete freedom in their work but are able to exercise political power. The professionals and administrators have a mix of freedom and responsibility appropriate to their positions. The administrative estate exercises a greater degree of power as compared to the professional estate, while the professionals have more independence as befits their role as the practitioners of applied science.

But even more significant for this neat division of responsible power and autonomous truth is the fact that the system comes equipped with a built-in set of checks and balances. The new Madisonian model takes up where the old one left off. The four estates are independent in some ways and mutually interdependent in others. Each protects its own interests with determined vigor, yet each must rely on the other three for its successful functioning. The estates must work together. They do in fact curb each other's excesses. And they must occasionally cross the recognized lines of functional separation to make sure that their ends are properly regarded by the others. "The scientists and professionals, in order to do their jobs," Price observes, "must be involved in the formulation of policy, and must be granted wide discretion in their own work."[59] "Politicians and administrators must control the key aspects of technological plans if they are to protect their own ability to make responsible decisions."[60] Price illustrates with a number of examples taken from the recent history of American government–science relations. An effective system of checks and balances has in fact been established, and a new "pluralist consensus" has grown up around them.[61] It could not have been better if it had been planned.

As we behold the new American Republic and marvel at its intricate equilibrium, we should note that Price's scholarship is by all accounts basically sound and his evidence accurate. *The Scientific Estate* is far and away the strongest work in the existing literature on science and politics and is fully in accord with the best research on the subject. From all reports, it is apparently useful to those who work in the circumstances the book describes. It helps members of the four estates to understand what they have been doing all of these years and to know their way about, especially their way about Washington, D.C. In this respect the book is of considerable value.

There are, nonetheless, some crucial problems in Price's model of politics. He wishes to show that we have been rescued from the rule of "faceless technocrats" by the fortuitous rise of a new pluralism.[62] In this way liberal representative democracy has been saved from possible destruction at the hands of the scientific revolution. But one wants to ask, Exactly who has been saved and how? What is included in this idea of political life and what is conspicuously missing? If Price's evaluations are at all correct, it would seem that liberal democracy has won, at best, a noticeably pyrrhic victory.

The strength of the new system, Price tells us, rests on its informality. It does not depend, as the Madisonian mechanism did, on precisely delimited powers and boundaries. If one were James Madison, one might worry over the fact that two of the estates, the professional and the scientific, exist completely outside the boundaries that the document of 1787 so carefully engineered. But Price assures us that all is well. "We have been establishing new kinds of checks and balances within the governmental system that depend on no legal provisions, but on a respect for scientific truth and professional expertise."[63] The new setup succeeds because the various actors and estates have different kinds of knowledge and respect each other's ways of knowing.

If this is so and if we are prepared to forget about legal safeguards on its account, a relevant question would still be whether the estates are equal in the kinds of knowledge they have. This is a critical matter when one thinks about the role of the political estate. What sort of special knowledge or expertise do the politicians command? Is there a distinctive political knowledge relevant to the new system? Traditional technocratic theories argued that politics had nothing of substance to contribute to a society based on expert scientific practice. As we saw in the *New Atlantis*, the political function was to become mere ritual gloss on the real business of governing the polity. If there are kinds of knowledge and knowledge holders more potent than others in the system Price describes, then the supposed informal checking and balancing might well go out of kilter.

The world Price describes runs almost exclusively on the knowledge of science and the facility of technology. The political estate is basically an occupational-functional group like the other three, but it is distinguished by the fact that it has nothing (or very little) to contribute to the realm of productive knowledge. If the politician wants to become relevant to the new order of things he or she must find out what the other estates are doing well enough to have a rank amateur's view of things. Much of the last part of the book tries to find adequate ways to keep the president and Congress "advised" in this regard.[64] Price sometimes speaks of the politicians as if they ought to be something like expert technicians of the political sphere. If this were true, it would no doubt earn the politicians the respect among equal colleagues upon which the unwritten constitution rests. But as Price makes abundantly clear, the whole political realm, especially as it must consider the new world of science, is characterized by a pervasive and embarrassing condition of ignorance.

Signs of Price's view here can be seen in his description of the ordinary citizen and his representatives. The common man, he states, used to be a radical force on the American political scene. He was a democrat, an activist, a reformer. But now things have changed. The true radicals in the nation are the scientists and professionals.[65] It is their work that brings about rapid technological change and social progress. They are the ones who have toppled the established order (quietly and politely) and built a new variety of politics. At the same time, the democratic radicalism of America's past has retreated into mass apathy. "The common man," Price writes, "is far less radical than the professions in what he demands of society; he is likely to prefer to be left alone in his traditional way of life, provided, of course, that technology can supply him with tranquilizers and television, the contemporary version of bread and circuses."[66] In a way that Spengler would have appreciated, the common man is dwarfed by the developments in scientific technics. He cannot participate in the governance of the new system because he is incapable of comprehending the information necessary to participate. His elected representatives are not much better off; they too are baffled by the mysteries of scientific society. The ordinary citizen, Price concludes, "cannot understand science, and is not in the habit of electing representatives who know any more about it than he does."[67]

In the end, the system of government Price describes is not one that encourages or expects citizen participation, not one that relies upon any effective representative process, not one that includes much of a role for traditional politics at all. His checks-and-balances argument, heralded as a way of salvaging representative democracy, is actually a way of showing why it is unnecessary to worry about such things. At the higher levels where the important policies are made and carried out, government takes place through interdependent, self-limiting collegia of expert knowledge holders. The political estate still plays a part—one part in four—in this process, serving as a vestigial remnant of a system of power and authority eroded by the winds of change. This estate, unlike the others, is responsible to the electorate. And a portion of its job is to speak for the public interest, for example, to indicate which tranquilizers and television programs are currently in favor. But Price is careful not to argue that good government is brought about by an aroused public working through diligent representatives. No; that would upset the delicate balance. The examples he gives do not show politicians striving to direct the products of science to ends that have arisen from debate and conflict in the public sphere. The picture is, instead, that of intimate interestate cooperation about matters that involve little disagreement. The role of the politician is to work closely with those who know what they are doing in their own specialized fields. While Price likes to call this a system of checks and balances, it is better thought of as a kind of friendly bargaining on the highest level among persons whose ends are basically in accord.

Price does not believe that this arrangement needs reform. When he does suggest improvements, they amount to a refinement of the communications among the estates. In particular, the president, his staff, and the Congress need to be better informed so that the political estate can better fulfill its collegial function. From the public he asks only that it understand and approve the new system of governance and not be concerned that things no longer work according to the traditional design. And even though his argument points to the fact that the general public has not the foggiest notion of the new constitution, he assures us that the public does approve of the new lines of jurisdiction.[68] Ignorance is the foundation of the new legitimacy.

If Madison were to examine this new arrangement, he would probably not be annoyed by the feeble character of participation and representation in the model. His design for a republic outlined in *The Federalist Papers* did not place much confidence in an active citizenry anyway. He might even congratulate Price for preserving at least that much of the founders' original vision of politics. It is possible, however, that Madison would be concerned with the safety of the kinds of divisions, checks, and balances that occur in his successor's theory. Are Price's safeguards sufficient to prevent the flagrant abuse of power or to neutralize the threat of "factions" "adverse to the rights of other citizens, or to the permanent and aggregate interests of the community," which Madison feared most?

A peculiarity of Price's discussion is that while it raises the problem of the fusion of economic and political power, nothing in the model of government he develops speaks to that issue. This fusion, remember, is the direct outgrowth of necessities of planning and coordination ushered in by scientific technology. Government and the large corporations must work hand in hand in order to finance and manage large-scale technological systems. Thus, the boundary between public and private breaks down (and Madison's brow begins to furrow). The thrust of Price's later argument leaves this condition alone. His forte is to unravel and set right the "diffusion of sovereignty" issue. He describes, explains, and justifies the relationships of power and authority existing between the four estates; his success is to illuminate this state of affairs for those actually involved in it and for those with an academic fascination with the matter. But the condition of "fusion" is taken as given and in the nature of things. After the second chapter of the book it is not subject to further scrutiny. His view is apparently that since "in the modern world there is no way to keep government and business from being dependent on each other," why worry about it?

A growing body of literature documents the concentration and astounding abuse of power in the high technology networks which bind together government agencies and certain large American corporations. The writings of H. L. Nieburg, Seymour Melman, Ralph Lapp, Richard Barnett, and others[69] identify a general condition in which the four estates—that is, congressional committees; government administrators; corporation managers, scientists, and engineers in and out of government; and others in the strata Price describes—work together to channel billions of tax dollars to military and aerospace research, development, and implementation projects which smack of narrow self-interest, self-perpetuation, and waste. In Nieburg's words, "The R & D cult is becoming a sheltered inner society isolated from the mainstream of national needs. More and more it departs from the reality principles of social accounting, insulated against realism by the nature of its contract relations with government and its political influence."[70] Far from being checked and balanced by its own internal workings, this system scrupulously avoids such limits and, in Nieburg's opinion, has the effect of "eroding the foundation of democratic pluralism."[71] "The open-end, cost-plus nature of the contract instrument, the lack of product specifications, official tolerance of spending overruns, all of which increase the total contract and fee (in a sense rewarding wasteful practices and unnecessary technical complication), permit violation of all rules of responsible control and make possible multiple tiers of hidden profits."[72]

I shall not dwell on these matters further but merely point out that Price begins his analysis of political power after a great deal of the available power has already been meted out and the possible problems associated with it simply dismissed. Price concedes a great deal to the megatechnical corporate-government alliance; *then* he begins his effort to rescue representative democracy. One cannot help feeling that in the political game Price describes, democracy plays only at halftime.

Echoes of his unfortunate approach recur throughout his analysis. For instance, he occasionally speaks as if the upper-level directors of business corporations are actually members of the *political* estate. Indeed, using the estate model, this is probably an accurate identification. Summarizing the full "division of power"[73] in the new system, he lists "the new estates of the scientific era, as well as branches of government and business corporations."[74] Price calls our attention to "a new kind of autonomy at many points within the vast complex of mixed government-business interests."[75] But he does not begin to criticize the nature of this autonomy.

Another important statement of a contemporary technocratic vision is found in the writings of John Kenneth Galbraith. Much more aware of the political implications of his argument than Price, Galbraith traces the rise of a newly emergent technical stratum located in the large, technically sophisticated corporations of our time. He is concerned that the existence of this stratum places in jeopardy the ability of modern society to guide its affairs through the application of widely held, positive social values. Where Veblen shifted the emphasis from entrepreneur to engineer and Burnham shifted it from engineer to manager, Galbraith insists that the focus of power and authority has moved once again. It no longer makes sense to talk about the rule of the managerial elite or of managerialism because the term *manager* takes in "only a small proportion of those who, as participants, contribute information to group decisions."[76] "This latter group," he explains, "is very large; it extends from the most senior officials of the corporation to where it meets, at the outer perimeter, the white and blue collar workers whose function is to conform more or less mechanically to instruction or routine. It embraces all who bring specialized knowledge, talent or experience to group decision-making. This, not the management, is the guiding intelligence—the brain—of the enterprise. There is no name for all who participate in the group decision-making or the organization which they form. I propose to call this organization the Technostructure."[77]

As I noted earlier, Galbraith holds that there are "imperatives of technology" that impose a definite shape upon modern industrial practice. Sophisticated technologies (1) increase the time separating the beginning of a task to its completion, (2) require great amounts of capital, (3) bring an inflexible commitment of resources, (4) require specialized manpower, and (5) require highly developed organization.[78] Navigation amid these imperatives is the métier of the technostructure. Foremost among the requirements of the industrial system, Galbraith argues, is the need for planning. "Conditions at the time of the completion of the whole task must be foreseen as must developments along the way. And steps must be taken to prevent, offset or otherwise neutralize the effect of adverse developments, and to insure that what is ultimately foreseen eventuates in fact."[79] The ability to plan, therefore, combines accurate foresight and adequate control. In the early history of industrial society, it was possible for the individual entrepreneur to handle his own information processing and planning. But under mid-twentieth-century conditions, that role can be successfully done only by technically trained collegia. Galbraith cites the example of Henry Ford in the 1940s as an illustration of the disastrous consequences that result from any attempt to retain autocratic, entrepreneurial control. Thus, an "organized intelligence" is the sine qua non of modern corporate planning and performance.

From his explanation linking technological imperatives, planning, and the necessity for *group* decision-making, Galbraith argues that the fundamental goal of the technostructure is to achieve autonomy. This, he asserts, is fully borne out in many observed cases in corporate life. The technostructure seeks to ensure its survival, protect its independence, and maintain control over those variables involved in its work. Both within the corporation and in all relevant areas of its economic and social environment, the technostructure moves to overcome all circumstances of interference, uncertainty, or limitation relevant to its activities. By maintaining an acceptable (but not maximum) level of profits for the firm, the technostructure guarantees its independence from the corporate owners, the diffuse and disorganized stockholders. Through a variety of means it controls both supply and demand and thereby eliminates the sordid perils of the market system.

Galbraith locates the basis of his autonomous technostructure in the autonomy of technical factors. Hence, persons involved in the quest for independence and control are not particularly power hungry. They harbor none of the base motives of tyrants or despots; very little of their thinking and behavior has anything to do with building a self-contained, self-conscious ruling class. Instead, their drive for power comes almost exclusively from a recognition of the necessities of effective performance in a world of advanced technology. They have seen the printout on the wall. If the technostructure of corporation *X* does not act creatively within the context set by technological imperatives, then a similar decision-making group in corporation *Y* or *Z* certainly will. The conduct of any particular instance of "organized intelligence" is simply that which any such intelligence, similarly prepared, would follow under the same circumstances.

Traditional kinds of economic competition are often excluded from this new arrangement. But competitiveness of a different style—the quest to devise the most effective arrangement of the given technical-economic ensemble—is still present. In this context, motives formerly associated with competition are radically altered. No longer do we find, according to Galbraith, capitalist entrepreneurs striving for ever higher profits or trying to force competitors out of business. The will of the technostructure consists of a strong desire to ensure that the organization succeeds, that individual positions in the organization are enhanced, that there is a respectable level of growth in corporate operations, and that progress be made in fine points of technical virtuosity.[80] Often this means that similarly trained persons in the technostructures of different firms will cooperate in ways that enhance the effective technical arrangement of a particular branch of industry as a whole, while denying the old-fashioned mechanisms of capitalist competition. This does not necessarily involve any unsavory collusion (although there are instances of overly zealous "planning," illegal price fixing, and the like). The image is, rather, that of technical soul brothers acting in the best interest of the corporation and, since corporations are presumably a reflection of the society they serve, the social interest as well. Such, according to Galbraith, is the technostructure's own conception of its role.

An inevitable concomitant of this ambitious quest for control is political power. The arms of the technostructure reach into the state, not by some selfish design but through the need to meet the requirements of the vast new technological system. "The mature corporation . . . ," Galbraith writes, "depends on the state for trained manpower, the regulation of aggregate demand, for stability in wages and prices. All are essential to the planning with which it replaces the market. The state, through military and other technical procurement, underwrites the corporation's largest capital commitments in its area of most advanced technology."[81] When the corporation cannot accomplish its ends through its own power, it relies on the power of government. This is made possible by the fact that there exists a rough harmony of interest between the state and the corporate system. Both want a stable economy, continued economic growth, a strong national defense, and continued scientific and technical advance. Government agencies work so closely with corporate organizations that the lines separating them often become indistinguishable. "The industrial system, in fact, is inextricably associated with the state. In notable respects the mature corporation is an arm of the state. And the state, in important matters, is an instrument of the industrial system."[82] By far the most convenient solution to the technostructure's problem of obtaining a large, steady supply of capital, relief from outside interference, and insurance against market uncertainty is simply to do business with the military. As was the case for Price, Galbraith's best examples come from the realm of "defense procurement."[83]

Galbraith has no intrinsic objections to the concentration of power in the "new industrial state." He is troubled, however, that the power will not be used to achieve a broad range of social goals. More precisely, he is bothered by a situation in which the values of an autonomous technostructure are becoming the only important values of the whole society. In this manner, the entire industrial-state network, what he now calls "the planning system," looms as a kind of self-programming mechanism, unresponsive to the wants of people it supposedly serves.[84] "If we continue to believe that the goals of the industrial system—the expansion of output, the companion increase in consumption, technological advance, the public images that sustain it—are coordinate with life, then all of our lives will be in the service of these goals. What is consistent with these ends we shall have or be allowed; all else will be off limits. Our wants will be managed in accordance with the needs of the industrial system; the policies of the state will be subject to similar influence; education will be adapted to industrial need; the disciplines required by the industrial system will be the conventional morality of the community. All other goals will be made to seem precious, unimportant or antisocial. We will be bound to the ends of the industrial system. The state will add its moral, and perhaps some of its legal, power to their enforcement."[85]

Galbraith offers evidence of this lamentable trend in the management of consumer demand, the circumvention of the market in matters of supply, and the cooperative disposition of political leaders. All sources of social direction outside the technostructure are gradually modified or neutralized. The hope for countervailing power, particularly that of organized labor against the corporation, emphasized in Galbraith's earlier writings, has now faded into oblivion. He now holds that labor unions have lost their punch and are thoroughly accommodated to the needs of the technostructure. Increasingly, therefore, the industrial system is able to establish its independence of the common-sense logic of technology that we examined earlier—the logic that runs from consciously chosen ends to appropriately selected means to the desired results. There is now, Galbraith argues, a "revised sequence." Those influences that should be leading—consumer preference, political decision, primary social goals—are actually dragged in train. Society begins to take its pattern from the warped priorities and projects of an economy of "unequal development" in which many worthwhile concerns—housing, health care, the arts, the welfare of the disadvantaged—languish in the shadow of the latest megatechnical investments.

But all is not lost, Galbraith believes. There is possible advantage in the fact that state and corporation are closely linked. "If the mature corporation is recognized to be part of the penumbra of the state, it will be more strongly in the service of social goals. It cannot plead its inherently private character or its subordination to the market as cover for the pursuit of different goals of particular interest to itself."[86] If only society would realize the true character of the existing circumstances and reassert its best purposes through the agency of the state, a manifestly bad system could be turned into a good one. In *Economics and the Public Purpose* Galbraith details an elaborate program of reforms. Among others, he suggests measures to revivify the market as against the planning system and to control government expenditures to accord with what he believes to be genuine public purposes. This would require an even more fundamental "emancipation of belief," freeing us from the idea "that the purposes of the planning system are those of the individual."[87] Accompanying this will be a wonderful rebirth of "public cognizance" in which the executive and legislative branches of government would regain their political fettle. From there one could move to a system of progressive taxation, regulation of prices and production, development of white-collar unions, and other steps to limit the influence of the technostructure.

The trouble with such proposals, revealed by Galbraith's own analysis, is that previously viable sources of political direction are missing. Each group with any appreciable social power has gained auxiliary membership in the technostructure or has been put on its payroll. Who, then, will speak for reasonable, humane, public ends? Where can any reformist tendency originate?

It is no exaggeration to say that at about the time Don Price picked up the idea of countervailing power, Galbraith was discarding it. Now, in an ironic (but obviously unconscious) swap, Galbraith embraces Price's notion of the scientific estate as the saving remnant. Social science is stranger than fiction. Citing Price's book as documentation,

Galbraith announces that only the intellectual community of the modern age—the scientific and educational estate in the university—can give voice to nontechnostructural values. "Unlike members of the technostructure, the educational scientific estate is not handicapped in political action by being accustomed to function only as part of an organization. It gains power in a socially complex society from its capacity for social invention."[88] "No intellectual, no artist, no educator, no scientist can allow himself the convenience of doubting his responsibility. For the goals that are now important *there are no other saviors.* In a scientifically exacting world scientists must assume responsibility for the consequences of science and technology."[89] He concludes that the industrial system has unwittingly brought into existence, "to serve its intellectual and scientific needs, the community that, hopefully, will reject its monopoly of social purpose."[90] The scientific and educational estate becomes the universal human class, carrying the banner of truth and virtue.

We recall that Don Price found it necessary to transcend this simple notion to seek a balance of power to keep the new estate honest. He recognized that scientists and professionals, including those in the university, have been dealt into the game, strive to protect their own interests, and cannot in themselves be counted upon for the critical independence necessary for reform. According to Galbraith's own arguments, the suddenly announced hope in the new intellectuals is ill founded. *The New Industrial State* shows clearly that members of the scientific and educational estate are deeply enmeshed in the projects and goals of the technostructure. He writes disparagingly of his own discipline, economics, saying that it "has extensively and rather subtly accommodated itself to the goals of the industrial system."[91] Other disciplines are just as strongly affected and tune their efforts to the kinds of inquiry that the industrial-state system will finance and utilize. The only exceptions that Galbraith finds are "the classics, humanities, some of the social sciences."[92] Oh, bright hope in yonder classics department!

In the end, Galbraith finds himself hoisted with his own petard. The argument that establishes the autonomy of the technostructure makes attempts to limit this autonomy seem flimsy and contrived. Late in the book he concludes, "The autonomy of the technostructure is, to repeat yet again, a functional necessity of the industrial system. But the goals this autonomy serves allow some range of choice."[93] Nothing in his position, other than the extraneous glimmer of the scientific and educational estate, shows how such choices might be made. Much like Price, he ends up saying that the new decision makers and planners of a technological society *must* be "responsible." "If other goals are strongly asserted, the industrial system will fall into its place as a detached and autonomous arm of the state, but responsible to the larger purposes of society."[94] But the question of how autonomy and responsibility can actually exist together in this kind of environment is not illuminated by further argument.

Galbraith's work is far from being the last word on American political life. It stands as an intelligent attempt to describe the technocratic strand and to come to grips with dilemmas this poses to liberal politics. But certainly there is more to the story: other bands are present in the spectrum of power. One problem with the technocratic perspective is that in order to make its case clear and significant, it must play down sources of business and financial influence, including those directly linked to technically trained elite groups, namely, the corporate owners. The tendency is to portray corporate ownership as entirely fragmented, withdrawn, content with its level of profits, and not at all eager to exercise immediate, day-to-day control. Power falls to the technically trained stratum in the corporation and society as a whole after the capitalist class has retired (or absconded). Just as Burnham found the holders of capitalist wealth "on yachts and beaches, in casinos and traveling among their many estates,"[95] Galbraith believes that corporate stockholders are so numerous, disorganized, yet satisfied that they could not possibly reassert their former position of supremacy. Is it possible, however, that the technostructure actually exercises power through the good graces of the financial and business elite and is directed in its work by tacit limits and understandings that reflect this influence? Although I cannot settle that question here, it is certainly worth asking.

The point is, then, that the writings of Galbraith and Price provide a partial, although important, portrait of the power relationships in modern American society. They describe a sphere of political and social influence which is already very large and gives every indication of continuing to expand. In this respect *The Scientific Estate* and *The New Industrial State* put flesh and bone on a tradition of technocratic theory that was previously little more than futuristic speculation. The major strength of both authors is simply to observe the mundane conditions now existing in the corporations, legislatures, universities, government agencies, and foundations and to report what they have seen. Earlier authors who approached technocracy through abstract theory or utopian fiction would not be surprised at the general shape of the world that Price and Galbraith describe. Veblen could survey these findings, notice a few new terms, and say, "Of course." Saint-Simon might complain that he had not been given sufficient credit for a basic outline put down a century and a half ago. And Bacon might express his appreciation at seeing the *New Atlantis* brought nicely up to date.

But the writings of Galbraith and Price are interesting not only for what they say about the role of the new men of knowledge but also for what they do not say. Clearly visible in both books is a new notion of political enfranchisement which neither author finds unusual or at all distasteful. The guiding assumption is that in a society based on sophisticated scientific technologies, the real voting will take place on a very high level of technical understanding. The voice that one has will depend directly upon the information, hard data, or theoretical insight one is able to supply in a group decision-making process. One may register to vote on this level only by exhibiting proper credentials as an expert. The balloting will be closed to the ignorant and to those whose knowledge is out of date or otherwise not directly relevant to the problem at hand. Among the disenfranchised in the arrangement are some previously formidable characters: the average citizen, the sovereign consumer, the small stockholder, and the home-grown politician. All of these have lost their vital role in the arena where the crucial policies and decisions are made. The historical evolution of technology with its accompanying demand for high-quality information has rendered their old-fashioned ways obsolete.

Neither *The Scientific Estate* nor *The New Industrial State* questions or criticizes this condition in the political order. Some things just have to be taken as "given" or inevitable, or so many of the leading writers in this field would have us believe. The solution Price offers the new polity is essentially a balancing mechanism, which contains those enfranchised at a high level of knowledgeability and forces them to cooperate with each other. Galbraith's cure holds out a virtuous elite within an elite to champion values lost in the new chambers of power. Beyond that, however, he will evidently be satisfied if the system yields "a reliable flow of income and product at reasonably stable prices."[96] But neither author does much to lend hope to the political possibilities of democracy or representative government under the new set of circumstances. Given the logic of their own arguments, it is probably unlikely that such possibilities could be sustained at all.

It is not unusual that liberal political theory should end up with elitist, nondemocratic conclusions. Such is the rule rather than the exception. What is interesting here, however, is that Galbraith and Price are so thoroughly imperturbable in the face of these developments. Perhaps a close familiarity with the relations of power and authority in liberal technocracy has convinced them that any serious effort to alter the situation would be foolish. An understanding of this kind can, very naturally, make one yearn for an even more advanced elite of the knowledgeable—a corps of enlightened planners, technology assessors, systems designers, or some other—to return the republic to an intelligent, humane course.

This does not end our treatment of the issue of technocracy. Having seen some of the problems it creates for liberal politics, we shall later want to examine its implications for socialism. This will be done, however, only after the basic theme has been substantially reworked and reinterpreted. In and of itself, the search for elites overlooks much of genuine importance. Larger and more interesting problems beckon us.

Chapter 5

Artifice and Order

Who governs? According to a leading opinion of our time, the question exhausts nearly all political possibilities. It is, indeed, the question addressed in the preceding chapter. But if one is interested in the relationship of technology and politics, the matter of "*who* governs" tends to skew the inquiry in a markedly lopsided direction. It suggests that we limit our search to persons and groups in positions of power and to compare the relative degrees of influence and control that such persons hold. Both liberal social scientists and Marxists, whatever their disagreements on the answer and its meaning, are agreed that "*Who?*" is the crucial question. I would certainly not dispute the fact that this is an important concern. But lost in the perspective is an equally interesting question: *What* governs?

Are there certain conditions, constraints, necessities, requirements, or imperatives effectively governing how an advanced technological society operates? Do such conditions predominate regardless of the specific character of the men who ostensibly hold power? This, it seems to me, is the most crucial problem raised by the conjunction of politics and technics. It is certainly the point at which the idea of autonomous technology has its broadest significance.

If one returns to the modern writings on technocracy in this light, one finds that parallel to the conceptions about scientific and technical elites and their power is a notion of order—a technological order— in which in a true sense no persons or groups rule at all. Individuals and elites are present, but their roles and actions conform so closely to the framework established by the structures and processes of the technical system that any claim to determination by human choice becomes purely illusory. In this way of looking at things, technology itself is seen to have a distinctly political form. The technological order built since the scientific revolution now encompasses, interpenetrates, and incorporates all of society. Its standards of operation are the rules men must obey. Within this comprehensive order, government becomes the business of recognizing what is necessary and efficient for the continued functioning and elaboration of large-scale systems and the rational implementation of their manifest requirements. Politics becomes the acting out of the technical hegemony. In the words of Herbert Marcuse. "Technological rationality reveals its political character as it becomes the great vehicle of better domination, creating a truly totalitarian universe in which society and nature, mind and body are kept in a state of permanent mobilization for the defense of this universe."[1]

Traces of this view occur in the very first modern writings on science and society. Saint-Simon recognized that if the design for an industrial society organized on scientific principles were successful, then the rule of his congress of industrialists, scientists, and artists would be subordinate to the rule of the principles themselves.[2] A truly well-governed society would, as a matter of course, adopt the laws of science and the methods of technology, which would in their perfection take care of every aspect of life. The new world would rest on a self-directing, self-maintaining system, which would require human presence only as a kind of lubrication in the joints. In twentieth-century social philosophy the conception of a self-maintaining technological society has recurred in a number of interesting and disturbing books—Oswald Spengler's *Man and Technics,* Friedrich Georg Juenger's *The Failure of Technology,* Karl Jasper's *Man in the Modern Age,* Lewis Mumford's *The Myth of the Machine,* Herbert Marcuse's *One-Dimensional Man,* Siegfried Giedion's *Mechanization Takes Command,* and Jacques Ellul's *The Technological Society.* These works contain a widely diverse collection of arguments and conclusions, but in them one finds a roughly shared notion of society and politics, a common set of observations, assumptions, modes of thinking and sense of the whole, which, I believe, unites them as an identifiable tradition. Taken together they express an inchoate theory which takes modern technology as its domain.[3]

My aim in this chapter and the next is to present a theory of technological society and politics based upon the writings mentioned, extending and clarifying arguments in that tradition whenever it seems useful. The approach will not be one of summarizing or comparing various kinds of theoretical formulations or of tracing their intellectual pedigrees. That Marcuse is similar to Ellul or Mumford in some ways, different in others, and that the differences have to do with their philosophical roots, are obvious conclusions and need not be agonized over for our purposes. Instead, I will examine and put to use the basic categories and assertions which are, in one form or another, crucial to any theory of this kind. What the perspective accounts for is not so much the full range of relationships present in any existing society but rather a tendency increasingly characteristic of the modern age. In describing this tendency the theory offers a picture of our world from a special vantage point, a picture that includes as a basic part of its structure a set of warnings about some of the destinations encountered on our society's present course.

The Technological Society: Groundwork

There is, unfortunately, a good deal of clutter along this path to understanding. Much of what now passes for incisive analysis is actually nothing more than elaborate landscape, impressionistic, futuristic razzle-dazzle spewing forth in an endless stream of paperback nonbooks, media extravaganzas, and global village publicity.

The images here have become common currency. We are now aware that whole continents are clogged with sprawling megalopolises linked together by circulatory systems of freeways and automobiles. We have become used to the presence of massive, labyrinthine bureaucracies manned by nameless, faceless functionaries and buoyed up by seas of forms, punch cards, and paperwork. We are, similarly, accustomed to hearing the facts of the production-consumption cycle: how much is manufactured, how many resources are used, how much is purchased, how much is ingested, how much is thrown away, and how much all of this has to do with the growth of large corporations, conglomerates, and multinationals. Every intelligent woman or man is now fully informed of the burgeoning, worldwide military systems combining the most spectacular and costly weapons with vast administrative organizations, all designed for surefire peace or annihilation, depending on one's point of view. We are familiar with the growing influence of electronic circuitry, communications networks, computers, and data banks upon the affairs of all institutions and individuals. And it is equally apparent that the world in which all of these are present and appropriate is one that brings an increasing use of wonder drugs for the body, mind, and spirit. As we become acclimated to such developments in their spectacular multiplicity, the future looms as less a shock than a bore.

To go on describing such things endlessly does little to advance our insight. Neither is it helpful to devise new names for the world produced. The Postindustrial Society? The Technetronic Society? The Posthistoric Society? The Active Society? In an unconscious parody of the ancient belief that he who knows God's secret name will have extraordinary powers, the idea seems to be that a stroke of nomenclature will bring light to the darkness. This does make for captivating book titles but little else.[4] The fashion, furthermore, is to exclaim in apparent horror at the incredible scenes unfolding before one's eyes and yet deep in one's heart relish the excitement and perversity of it all. Alleged critiques turn out to be elaborate advertisements for the situations they ostensibly abhor.[5] Exceptions to this sad state are to be found in a handful of serious artists and philosophers whose efforts go beyond anecdote and future prattle to penetrate the aesthetic or theoretic essence of the phenomena at hand.[6]

Even those who launch serious investigations, however, frequently encounter snags along the way. This is understandable. The subject, for reasons discussed earlier, is an overwhelming one in its extent and diversity. It is characteristic of those who have recognized technology as a problem for politics that they accepted the challenge without any great love or eagerness. In most cases the thinker seems to have been driven to accept the technological perspective by the weight of circumstances which could no longer be denied. For writers like Jaspers, Giedion, Ellul, Arendt, and Heidegger, the facts of technics loom as a colossal barrier that the modern intellect must penetrate even though it would prefer to move on to something more agreeable. Responding to the staggering proportions of the problem, thinkers have gravitated toward Scylla on the one hand, Charybdis on the other—between an altogether crass reductionism and a wild, multifactoral confusion.

Most notorious of the reductionists is Ellul. His work stands as an elaborate hall of mirrors, deliberately designed to leave no passage out. From the building of skyscrapers to roller skating, Ellul's ubiquitous concept of technique expands to encompass any subject and to resist contrary examples. Planning and public opinion, for instance, which might be taken as possible forces counter to unbridled technical advance, are shown to be mere products of technique itself.[7] *The Technological Society* is less an attempt at systematic theory than a wholesale catalog of assertions and illustrations buzzing around a particular point. The book is one that its readers rave against, refute in dozens of ways, and then lose sleep over. Fortunately we need not follow Ellul in seeing technique as a strictly univocal phenomenon or in making totally nonfalsifiable arguments. It is possible to learn from the man's remarkable vision without adopting the idiosyncrasies of his work.

When writing on the technological society does not bog down in reductionism, it frequently moves toward the opposite extreme and tries to deal with all varieties of explanation simultaneously. Analyses of this sort become sprawling grab bags of sociological information about the expansion of population, market conditions, mass culture, and the psychological condition of modern man. There is nothing inherently wrong in this. When it comes right down to it, most modern thought amounts to combining theories and conclusions whose logical connections are indefinite. We tend to believe that things happen as the result of "a little of this and a little of that." Even the determinism of Karl Marx was modified to include a rich variety of historical factors whenever Marx wished to interpret a specific historical case, for example, the movement from feudalism to capitalism or the workings of French politics.[8] The problem with allowing the inevitable flood of variables to enter a theory too early, however, is that one may fail to treat the original subject of concern in its integrity.

What we require, then, is to give the technological element its due, without falling into either reductionism or eclectic compromise. If it makes sense to isolate this subject for attention, the perspective based on it should give us something distinctive and useful. It should show how it is that if one pulls on this particular string, one uncovers a section of the edifice not previously visible.

In this light I want to set forth some concepts and distinctions that seem to me central to a critical theory of technological society.

Artificiality

An obvious place to begin is to recognize that the human world has entered a new stage in its history, a stage qualitatively different from anything previously known. Perhaps the most important fact of this new age is that in a very literal sense men and women are constructing the world anew. Science has revealed the structures, processes, and laws of nature and thereby opened the natural sphere to all sorts of modifications. The material world in which man now lives is, as a result, one of artificial resynthesized products. The original creation was not sufficient to suit Western man's striving, appetitive constitution. A second creation has taken place in which human beings refashioned vast portions of material reality to suit their desires.[9]

The scope of artifice and modification, furthermore, does not stop with the reconstruction of nature. It also extends to human society. Much of what had previously rested upon "tradition" or "natural social groups" has now been rebuilt according to preconceived design. Roles, relationships, groups, and even individual personalities are now largely subject to conscious technical manipulation.[10]

A consequence of artificiality is that human beings find themselves responsible for an increasingly large share of worldly conditions. Structures of a natural or traditional sort were for the most part self-maintaining in the sense that deliberate control was not required to keep them intact. Artificial structures, in contrast, must be man-maintained since the "second nature" now produced is not a part of the world's original process of self-adjustment.[11]

Extension

Beginning with Rousseau's *Discourses* and continuing to Marshall McLuhan's writings on electronic media, modern writers have observed that technical devices in a real sense extend human capabilities.[12] The five senses, physical mobility, human strength, and ability to act are all augmented and, in some points of view, enhanced by such means. One may even classify technologies in terms of which human capacity is extended and how. Transportation technologies increase one's ability to move in geographical space. Systems of communication expand one's power to see, hear, speak, and express oneself over great distances. Technology in this light is significant for the fact that it allows one to do things one could not previously have done. Perhaps more accurately, it allows human individuals to *be* something, a kind of being, previously unknown on earth.

A consequence of extension is that, in a way never experienced before, a person's physical presence is not required for action to take place. Every variety of control, expression, thought, motion, and production of events can take place through remote channels composed of long lines. Individuals can substitute "being there" with a number of sophisticated devices. A by-product of extension, therefore, is a quality of remoteness.

Rationality

This is a concept indispensable to any adequate description of technological society, but there is far from universal agreement on its meaning. The notion can be understood to indicate (1) a state of intellect, (2) a way of acting, (3) an arrangement of things in the world, or all these together. Michael Oakeshott has added to the controversy by suggesting that any addition of the technical element to the meaning of *rational* distorts the original meaning of the term (reasonableness or intelligence in the conduct of a given activity) and tends to corrupt the practice it identifies.[13] But allowing for this complication, the following uses of the concept are fairly standard.

Rationality is a condition of reasonable or logical order in either the static structure of any instrument or in the dynamic operation of any technique. The parts fit together correctly. Steps in the process follow each other in an appropriate, preconceived fashion. Rationality in this sense is something that can be modeled in a formal design or programmed in formal rules of operation.[14] In Ellul's version, "Every intervention of technique is, in effect, a reduction of facts, forces, phenomena, means, and instruments to the schema of logic."[15] It was an ordering or "systematic arrangement" of this sort that Max Weber identified as a linear process of transformation in all social history. In this process, static designs of intellect gradually conquer the charismatic, nonrational elements in history and come to dominate all of human existence. "The fate of our times," he concluded, "is characterized by rationalization and intellectualization and, above all by the 'disenchantment of the world.'"[16]

Another definition of rationality, which we have already seen, is that of "the accommodation of means to ends."[17] Since problems connected to this theme will concern us later, I shall say nothing more about it now.

Rationality can also be understood as a kind of efficiency. Something is rational if it produces the desired effect with a minimum of waste or, to put it differently, if it gives the highest output per unit input. Organization theorist James D. Thompson identifies this as the economic criterion of technical rationality. "The economic question in essence is whether the results are obtained with the least necessary expenditure of resources, and for this there is no absolute standard."[18] It is this notion of rationality that permits Ellul to argue that, in point of fact, economics is now very largely subordinate to technique. Technological society began with economic calculations of the input-output ratio seen mainly in terms of profit and market performance. Now, Ellul argues, this standard of performance has become a much more general social norm, applicable in virtually every sphere of social activity. Profit as an economic motive and the market as a mechanism of social adjustment are both still present, but they have withered in significance through the universal application of input-output technical rationality—the *quest* for absolute efficiency, "the one best way."[19] It is this version of the idea of rationality that also allows Ellul and others to conclude that precise measurement, calculation, and rationality are merely different names for the same general process.

One could ferret out further definitions, but this would risk creating an unnecessary public nuance. Suffice it to say that there are at least three standard, distinct uses for the term "rationality" which make good sense. It is interesting to notice that they are in no sense necessarily compatible. An institution, policy, or practice may be rational in one meaning of the term and yet totally irrational by either or both of the others.

Size and Concentration

In their very nature, modern technologies are large-scale, high-energy, high-resource systems requiring massive commitments of capital and technically trained manpower. Small-scale, localized arts and crafts live on, of course, primarily as vestigial remnants of a tradition of material culture that has lost its vitality. While technical organizations and apparatus of enormous size are not a totally new phenomenon in history, they are clearly more central to contemporary social existence than to any previous era.[20]

Two general circumstances account for this quality. In many instances the fact of scale arises as a necessity of technical operation. If the system in question is to work *at all,* vast numbers of parts must be drawn together (concentrated) and arranged in a functioning whole. Under many present situations, the "tool" and its utility can only take shape when the aggregate of parts has reached a certain critical mass. In this sense our tools are simply much larger, far-flung devices than have ever been used previously. As technical goals have expanded in ambitiousness, so have technical systems expanded in size.

The impetus to large size is also linked to the input-output rationality mentioned above. Within certain limits, it is possible to rationalize an operation—that is, to increase efficiency or reduce costs—by increasing its size. Economies of scale in production, for example, are frequently gained by building more and larger factories or by coordinating previously separate activities in the same general field. If one adds to this the element of power that accompanies the elaboration of large-scale networks, power over the environment of operations, and corresponding freedom to act, the rationale becomes even more compelling. Enlarge, concentrate, and connect. From water systems to motel chains, telephone companies to ready-in-a-jiffy hamburger restaurants, the tendency is to arrange things on a vast scale.[21]

Of course one can argue that there is no inherent technical reason why even such ambitious activities as steel and automobile production, electric power generation, and distribution could not be done on a localized, modest scale. But this simply ignores the indelible link between the technical and the rational upon which our age has fastened. It is always possible in principle, perhaps even in some form of practice, to have a conveniently small aluminum plant cooking in the vacant lot next door. But, as analysts of scale are quick to point out, the costs of proceeding in this manner are enormous as compared to a mammoth network to handle the same function.

In the twentieth century, to propose that any socially important technology be organized on a small scale is usually taken as a sign of madness. A quaint, old-fashioned madness, but madness nonetheless. Enterprises of small scale are impossible given what must be done and totally irrational by the best input-output calculations. That conclusion is crucial to the perspective we are developing here. To begin seeing other possibilities involves a level of questioning and criticism far beyond anything advanced industrial societies are willing to undertake.

Division

Technology succeeds by taking the world apart and putting it back together in productive ways.[22] In the design of this success, precise division is absolutely crucial. Francis Bacon long ago stressed the idea that the laws and qualities of nature useful to man could be found only by cutting the universe to pieces and inducing each section to give up its secrets. "For I am building in the human understanding," he wrote, "a true model of the world, such as it is in fact, . . . a thing which cannot be done without a very diligent dissection and anatomy of the world."[23]

In the operation of technical systems the components must, at the outset, be separate and precisely defined. If the parts are nonhuman, they must be built to perform a specific, predetermined function. The idea of a division of labor is merely one way of seeing how division and specificity produce results. One might say that the division of labor is the basis of the industrial revolution; the division of all reality is the foundation of the technological revolution. Siegfried Giedion speaks of the process of division in terms of the machine model and arrives at a standard conclusion: "Mechanizing production means dissecting work into its component operations—a fact that has not changed since Adam Smith outlined the principle of mechanization in a famous passage of his *Wealth of Nations* in 1776. 'The invention of all those machines by which labor is so much facilitated and abridged seems to have been originally owing to the division of labor.' It need only be added that in manufacturing complex products such as the automobile, this division goes together with a re-assembly."[24]

Complex Interconnection

Precise division, then, is followed by intricate reconnection. The effective working of any technical apparatus or organization requires that large numbers of parts be linked together in rational, functional wholes. In the systems created, the operating parts have multiple and diverse interconnections. Increasingly in modern writing, one finds the noun *complexity* used to suggest a condition of this sort which might be studied in its own right. Complexity looms as a distinctive problem when systems of interconnected parts begin to tax the human ability to make the artificial whole intelligible.

Nature, of course, also contains complex interconnections. But as noted earlier, the complexity of natural structures does not pose difficulties for us (except when we wish to inquire into them or divide and rebuild them according to our own specifications). The structures of nature are "God given," self-regulating, and self-maintaining. Artificially complex structures, in comparison, pose a number of practical and intellectual problems; man is responsible for their synthesis, regulation, and continued maintenance. In almost no instance can artificial-rational systems be built and left alone. They require continued attention, rebuilding, and repair. Eternal vigilance is the price of artificial complexity.

Dependence and Interdependence

Often one wants to say more about the parts of a technological ensemble than that they are divided in their functions and interconnected. A useful next step comes in noticing that many of the most important interconnected components have a relationship of interdependency. The performance of technological systems rests upon the ordered and effective contribution of parts that rely on each other. Nothing of significance is done by self-contained units acting alone. Virtually everything is accomplished through the coordinated work of a variety of operating segments.

Care must be taken, however, not to draw absurd conclusions from this notion. There is a tendency to think that in an increasingly interdependent technological society or world system, all of the parts need each other equally. Seen as a characteristic of modern social relationships, this is sometimes upheld as a wonderfully fortuitous by-product of the rise of advanced technics. The necessary web of mutual dependency binds individuals and social groups closer together; lo, a new kind of community is forming before our very eyes. But this view involves distortion. It confuses interdependency with mere dependency. An individual may depend upon the electricity or telephone company for services crucial to his way of living. But does it make sense to say that the companies depend on that particular individual? It is hard to sustain the notion of mutuality when one of the parties could be cut off the relationship and the other scarcely notice it.[25] Not every plug and not every socket is essential to the network.

A completely interdependent technological society would be one without hierarchy or class. But the distinction between dependence and interdependence points to a hierarchical arrangement of the segments of the technological order, an arrangement that includes social components. Within each functioning system some parts are more crucial than others. Components that handle the planning or steering for the whole system are more central than those that take care of some small aspect of a technical subroutine. One may ask: On what does everyone and everything depend? On what do many things depend? Relatively few things depend? Rather than a condition of equality or a classless state, we find arrangements of subordinate and superordinate units. Large man-machine networks require hierarchical structure as a normal operating condition.

The Center

The necessity for coordination of diverse operations within a large-scale network usually requires central control. To say that something is centralized means that it has undergone a process that makes its working depend upon directions from a core. The distinction between the center and the other parts of a system can be of either a geographical or functional sort. Hence, centralization entails that control over basic technical means—energy, communication, water supply, sewage disposal, and so forth—is removed from geographical localities, for example, neighborhoods or local communities, and given over to a single agency of control. Under the input-output norms that characterize twentieth-century thinking, local direction of anything of genuine importance is seen as a source of waste, chaos, or worse. Coordination by a center can also be a purely functional matter. Relatively peripheral parts and persons in an organized hierarchy depend upon a core that makes plans and issues orders.

In most modern technological systems autonomous action is truly available to the center alone. Some have argued that peripheral parts also exercise a measure of control by selecting which information actually reaches the center. A few varieties of highly technical organization, think tanks like RAND for example, have found that attempts at central coordination are more of a hindrance than help in getting work done. Taking such paltry signs as a cue, some observers have begun to predict that a collegial decentralization is the wave of the future. But compared to the success of central control in likes of the Apollo program or the worldwide operations of ITT under Harold Geneen, the alleged counterexamples are pathetic.

Apraxia

To this point I have deliberately avoided neologism in developing this set of terms. But from the viewpoint I have sketched, one situation deserves a special name. If a significant link in a technical system ceases to function, the whole system stops or is thrown into chaos. It is this condition I want to call *apraxia,* a term used in medicine to describe the inability to perform coordinated movements. In large-scale technical networks composed of artificial components with complex interconnections and interdependencies, apraxia is a constant danger.

The possibility of severe breakdowns in high technology has been an important concern since Marx described the unwieldiness of capitalism in its advanced stages. But it was Thorstein Veblen who first stated the issue in a way similar to that I am using here. "The industrial system," he wrote, "runs on as an inclusive organization of many diverse, interlocking mechanical processes, interdependent and balanced among themselves in such a way that the due working of any part of it is conditioned on the due working of all the rest."[26] "With every further advance in the way of specialization and standardization, in point of kind, quantity, quality, and time, the tolerance of the system as a whole under any strategic maladjustment grows continually narrower."[27] Veblen saw the problem in terms of the "sabotage" of the industrial system by the industrialists and financiers. "The date may not be far distant," he warned, "when the interlocking processes of the industrial system shall have become so closely interdependent and so delicately balanced that even the ordinary modicum of sabotage involved in the conduct of business as usual will bring the whole to a fatal collapse."[28]

The threat of technological apraxia is an important cutting edge of the theory. It is a condition that planners and administrators regard with worshipful awe and critics sense to be the ultimate barrier to any attempt to arrange things differently. The technological order is one in which all systems are "go" and indeed must be. The alternative is disaster for technology-dependent human population. Whether through accident, as in the great East Coast blackout of 1965, or deliberate intervention, as was the case in the energy crisis following the Middle East war of 1973, the lessons here are sobering. For those who respect the possibility of calamitous breakdowns, there comes a moral imperative that all major systems must be kept in good functioning order. Any nonexpert tampering is seen as positively malicious. In visions of technological society, apraxia assumes much the same place as the return to the state of nature in Hobbes. It is the ultimate horror, a condition to be avoided at all costs.

These, then, are some of the terms I consider useful in thinking about the composition of a technological society. The reader may have noticed that one or more of his or her own favorite categories of description—organization, integration, control, uncertainty, and so forth—are not present in this array. Some of this lack I hope to remedy in the following pages as the syntax of this vocabulary becomes clear. We turn now to a core issue, the thematic axis about which the more specific questions in this perspective revolve.

Technological Politics: Master and Slave Revisited

We have already seen the manner in which the master-slave metaphor and the idea of absolute mastery stand at the heart of Western thought about science and technics. In the theory of technological politics the theme returns with a series of ironic twists. If there is a central conclusion shared by critics of the technological society, it is that man overcomes his bondage to economic necessity only by submitting to bondage of a different, but equally powerful, sort. The conquest of nature is achieved at a considerable price—an even more thorough conquest of all human and all social possibilities. The dream of scientific technics promises a boon that was long thought impossible: inexhaustible riches combined with a liberation from toil. Ellul compares this dream to the legend of Faust and concludes that in his contract with *la technique,* man did not read the fine print. "Man never asks himself what he will have to pay for his power," he observes. 'This is the question we ought to be asking."[29] The answer, Ellul makes evident on every page, is clear. The true price is loss of freedom.

Marcuse, in a similar fashion, again and again returns to the question, Why has human liberation been so long postponed? Technical mastery in the supreme form of automation ought to have reversed "the relation of free time and working time on which the established civilization rests: the possibility of working time becoming marginal, and free time becoming full time."[30] But this possibility has never been realized. Paradoxically, the very means to emancipation have instituted a new condition of servitude. "The modes of domination have changed: they have become increasingly technological, productive, and even beneficial; consequently, in the most advanced areas of industrial society, the people have been co-ordinated and reconciled with the system of domination to an unprecedented degree."[31]

Perhaps it is only a coincidence that at the same time that the artificial slave became a distinct possibility, Western philosophers began to have misgivings about the stability of the institution of slavery. In the famous passage on "Lord and Bondsman" in Hegel's *Phenomenology* we discover that the mode of absolute mastery is ambiguous and ultimately self-subverting. Hegel's treatment takes slavery to be one moment in the progress of Spirit toward full self-consciousness. The master tries to obtain recognition of himself by totally imposing his will on another subject. But the attempt does not succeed. "Just where the master has effectively achieved lordship, he really finds that something has come about quite different from an independent consciousness. It is not an independent, but rather a dependent consciousness that he has achieved. He is thus not assured of self-existence as his truth."[32] Master and slave are reciprocally defined, but, paradoxically, the slave has gained the upper hand, for the condition of the master's independence is a dependency of a radical sort: dependence on the slave himself. But the dialectical irony does not end here. The slave takes yet another step beyond the circumstance of his lord by laboring in the material world. In this way his selfhood takes on objective reality and substance, something the master totally lacks. "This consciousness that toils and serves accordingly attains by this means the direct apprehension of that independent being as its self."[33]

It was from passages like these in Hegel that Marx began thinking about the universally exploited proletariat and the relationship of wage slavery to the capitalist master class. Marx's theory restates the Hegelian irony in which subjugation contains the seeds of its own transcendence. In his version, of course, the slaves are able to act out their independence through revolution.

Another statement of an unexpected link between slave and master came in the late nineteenth century in the philosophy of Friedrich Nietzsche. Through a study of ancient etymology, Nietzsche claimed that he had discovered evidence of a "slave revolt in morals" which had occurred two thousand years earlier and had shaped civilized life up to and including his own time. The weakest and most sickly of human beings had in their craft and cunning subverted the power of the strong, noble men who had previously ruled supreme. This was accomplished by subtle invention in the meanings of certain terms of evaluation, particularly "good and evil" and "true and false."[34] Because the strong accepted the terms of their slaves, they could no longer exercise mastery and fell into decadence and submission.

The nineteenth century was, of course, the period in which slavery and serfdom were finally abolished worldwide. Our civilization, we sometimes forget, has gone just barely one hundred years without this venerable institution. And at about the time of the abolition, philosophers had begun to uncover certain paradoxical characteristics inherent in the institution itself. What is interesting for our purposes is that in slightly altered form the philosophical critique of slavery is recapitulated in modern speculation about the social context of technology. To cite an explicit case, the first use of the word "robot" in our language was directly linked to an idea of slave rebellion. Our concept of "robot" derives from the Czech *robota,* meaning compulsory labor.[35] The word was introduced into the English language by the popular play *R. U. R. (Rossum's Universal Robots)* by Karel Čapek, in which a newly invented working class of automatons rise in revolt against their human lords. The robot Radius announces to the head of the factory, "You will work! You will build for us! You will serve us! . . . Robots of the world! The power of man has fallen! A new world has arisen. The Rule of the Robots! March!"[36] The overt revolt of the artificial proletariat can be taken as a metaphor for a much more subtle kind of technical rebellion, which Čapek and others have tried to depict for their audiences. In the theory of technological politics, the conquest of humanity by technique more clearly resembles the ambiguous situations of slavery described by Hegel and Nietzsche than the revolutionary apocalypse of Marx.

These themes first set forth in metaphor and abstract symbolism can, however, be stated in a more precise, concrete fashion. As they occur repeatedly in modern European and American writings, the major assertions of the master-slave paradox in terms of modern technology are the following.

—that men have assumed a position of extreme and even pathological dependence upon their technical artifacts;
—that the adoption of complex technical forms brings with it a discipline upon individuals and society much more stringent and demanding than any other corresponding arrangement in history;
—that technical means in the context of social practice tend to become ends in themselves or to redefine the established ends to suit the requirements and character of their own operation;
—that the presence of sophisticated technologies in society tends to transform and dominate the mental habits, motives, personality, and behavior of all persons in that society;
—that technical artifice as an aggregate phenomenon dwarfs human consciousness and makes unintelligible the systems that people supposedly manipulate and control; by this tendency to exceed human grasp and yet to operate successfully according to its own internal makeup, technology is a total phenomenon which constitutes a "second nature" far exceeding any desires or expectations for the particular components.

Along with the notions of technological change and technocracy we have already discussed, these are the major conclusions advanced in the literature on autonomous technology in the twentieth century. Most of what is truly interesting or unusual in the theory of technological politics, I would argue, is based on an elaboration of these ideas.

The reader, then, has a set of general concepts in mind and a number of central themes in full view. My purpose from here on is to clarify and demystify the problems that these terms reveal. As we probe contexts in which the above conclusions make sense, we will find ourselves in full encounter with the questions of knowledge, control, and technical neutrality raised at the beginning of the book.

Order, Discipline, and Pace

It is important to notice, first of all, the conception of society which takes shape in the technological perspective. Absolutely fundamental is the view that modern technology is a way of organizing the world and that, potentially, there is no limit to the extent of this organization. In the end, literally everything within human reach can or will be rebuilt, resynthesized, reconstructed, and incorporated into the system of technical instrumentality. In this all-encompassing arrangement, human society—the total range of relationships among persons—is one segment. "Technological society" is actually a subsystem of something much larger, the technological order. Social relationships are merely one sort of connection. Individuals and social groups are merely one variety of component. The connections and groupings of inanimate parts are equally crucial to the functioning of the whole.

This is not to say that any existing society has been integrated in all its parts into a purely technological order. There are some kinds of social relationships, those involving love and friendship for example, that have not yet been fully adapted to the demands of technical routine. The position of the theory is that a strong tendency toward order of this kind is highly pronounced in all spheres of Western society and that its development will in all likelihood proceed rapidly on a worldwide scale.

An apt comparison can be found in the notions of order and society of the medieval Christian world view. In the great chain of being, with its hierarchy of God created things, men and women in their various social positions occupied merely one of several levels and by no means the most important stratum at that. Each being had its own "degree" or grade of perfection, and it was mandatory that each level in the hierarchy keep its established place. For humans to aspire to anything more than their appointed position was an act of sinful pride and defiance, an invitation to chaos.[37] In the present view, then, the liberatory quality of technology must be weighed against what its system of order imposes upon and requires of man. Attempts to deal with this side of the technical situation have engendered two distinctive themes in modern social philosophy: first, the mechanization of human activity and social relationships; second, the more thorough conditioning of individuals through their contact with technical systems and apparatus.

The idea of a mechanized humanity was a prominent part of nineteenth-century literature in Europe and America. Thinkers looked at the advance of an advanced, industrialized world and wondered openly about the capacity of man to retain his integrity in the face of such marvelous instrumentation. Thomas Carlyle, to whom we owe the idea that "man is a tool using animal. . . . Without tools he is nothing, with tools he is all," wrote of the possibility that men would internalize the external reality of mechanization and become themselves thoroughly mechanical in thought and behavior.[38] Predictions of this sort were not uncommon. The scientific world view of the nineteenth century was still centered in the Newtonian vision of a universe running like a colossal clock.[39] Since this model had served well in the development of physics and since the industrial machine had indeed proven to be an astounding force, the image of man as a machine was widely thought a natural. For some thinkers this meant that homo sapiens was quite literally *l'homme machine* of the sort described by the eighteenth-century philosopher Julien Offray de La Mettrie: "The human body is a machine which winds its own springs. It is the living image of perpetual movement."[40] In a totally mechanical universe, a mechanical man is an appropriate microcosm. To this day, views of this kind live on in the writings of Skinner, Woolridge, and others, accompanied by the peculiar conviction that if human beings are *in any way* like machines, then it follows that they either do or very well ought to behave totally like mechanical devices.[41] In the nineteenth century many of those who took violent objection to this conception of humanity were convinced that the masses of men had already become greatly mechanized in their personal and social existence. Man's relationship to the industrial process, particularly to the machinery, organization, and techniques of the factory, would eventually bring La Mettrie's prophecy to fruition in a remade human environment.[42]

The mid-twentieth century has brought the eclipse of the machine as a model for everything under the sun. Too many recent developments in science and technology—quantum physics, relativity, modern chemistry and biology, the alloys, plastics, the transistor—simply do not match the two primary images of the older mechanical tradition: Newton's clockwork universe and the cog and wheel machine of nineteenth-century industry. Artifice has become more subtle. Many devices properly called machines are no longer truly mechanical. Even Lewis Mumford, who emphasizes the idea of society as machine, has changed his emphasis to something called the "Power Complex." What needs expression is the idea of a set of large-scale, complex, interdependent, functioning networks which form the basis of modern life; for this, "the machine" will no longer suffice.[43]

But the decline of a metaphor does not mean that concerns it represented vanish. The possibility that man faces an unwitting bondage in his relationships with technical systems is still a living hypothesis. Father of this question in modern social philosophy was Jean Jacques Rousseau. In the *Discourses* Rousseau placed himself in open disagreement with the prevailing opinion of his time by arguing that the advance of the "arts and sciences" was a degenerative rather than a progressive movement in history. Human existence, he believed, had long ago reached something of a golden age in which men were free and happy. But then there occurred a "fatal accident"[44] (a chance discovery or invention?) that brought a great revolution in metallurgy and agriculture and thereby "civilized men and ruined the human race."[45] "As soon as some men were needed to smelt and forge iron, other men were needed to feed them. The more the number of workers was multiplied, the fewer hands were engaged in furnishing the common substance . . . and since some needed foodstuffs in exchange for their iron, others finally found the secret of using iron in order to multiply foodstuffs. From this arose husbandry and agriculture. . . . From the cultivation of land, its division necessarily followed; and from property once recognized, the first rules of justice."[46] Rousseau held that by adapting their lives to this early technological revolution and to subsequent ones, men had given up their original freedom and entered the enslaving web of dependencies involved in a complex economic society.

A more familiar conception of how freedom is lost emphasizes the presence of external restraints. The activity of an individual or group is limited by an outside factor, such as the presence of a stronger force or the restrictions of law. Much of liberal political philosophy attempts to delineate the conditions under which such restraints are or are not justified. The theory of technological politics, however, follows Rousseau in seeing the loss of freedom in the modern world as preeminently a situation in which individuals become caught up in webs of relationships which have a pathological completeness. Conditions of life for all persons come to be inextricably tied to systems of transportation, communication, material production, energy, and food supply for which there are no readily available alternatives. An automobile or mass transit network carries the person to work at a particular time every day. The food he or she eats is grown by agribusiness concerns and shipped in from a great distance and distributed by large chain supermarkets. Information about the world is available, prepackaged like the food one eats, in television news programmed at a central source. For all manner of day-to-day activities, apparatus like the telephone become absolutely indispensable. Because in most instances the working of such systems does not include the application of restraints, the problem of human freedom as it involves extreme dependency and helplessness seldom comes up.

Closely tied to the phenomenon of dependency is the situation of servitude within technological relationships. The human encounter with artificial means cannot be summarized solely (or even primarily) as a matter of "use." One must notice that certain kinds of regularized service must be rendered to an instrument before it has any utility at all.

One must be aware of the patterns of behavior demanded of the individual or of society in order to accommodate the instrument within the life process. There are, to put it differently, subtle but important costs as well as obvious benefits. These costs, usually forgotten or thought "inevitable" by those who must bear them, are in the aggregate truly staggering.

An early analysis of such circumstances in microcosm appears in Ralph Waldo Emerson's *Works and Days*. "Many facts concur to show," he says, "that we must look deeper for our salvation than to steam, photographs, balloons, or astronomy. These tools have some questionable properties. They are reagents. Machinery is aggressive. The weaver becomes a web, the machinist a machine. If you do not use the tools, they use you."[47] Although Emerson employs the image of the machine, his main point is not actually a standard "mechanization of man" thesis. It is instead the idea that all tools are "reagents"; they are not a passive presence in a human situation but instead evoke a necessary reaction from the person using them.[48] Attachment to apparatus not only requires that men behave in certain ways, it also gives them a positive responsibility and criterion of performance they must meet. "A man builds a fine house; and now he has a master, and a task for life: he is to furnish, watch, show it and keep it in repair, the rest of his days."[49]

Bruno Bettelheim offers a similar report from recent twentieth-century experience. "In my daily work with psychotic children, and in my efforts to create an institutional setting that will induce them to return to sanity, I have come face to face with this problem of how to take best advantage of all the conveniences of a technological age, . . . and to do it without entering a bondage to science and technology."[50] Bettelheim goes on to describe how he learned to study the shaping effect that any new device had upon the work of his institution. "Whenever we introduced a new technological convenience, we had to examine its place in the life of our institution most carefully. The advantages we could enjoy from any new machine were always quite obvious; the bondage we entered by using it was much harder to assess, and much more elusive. Often we were unaware of its negative effects until after long use. By then we had come to rely on it so much, that small disadvantages that came with the use of any one contrivance seemed too trivial to warrant giving it up, or to change the pattern we had fallen into by using it. Nevertheless, when combined with the many other small disadvantages of all the other devices, it added up to a significant and undesirable change in the pattern of our life and work."[51]

Bettelheim's statement reflects a level of awareness altogether rare in contemporary writing, not to mention the practices of everyday life. His eminently sensible conclusion, therefore, smacks of a certain radicalism: "The most careful thinking and planning is needed to enjoy the good use of any technical contrivance without paying a price for it in human freedom."[52] Such sentiments are generally thought to be antiprogressive and are ignored in most polite company. Rousseau's views, similarly, are widely believed to have been those of a romantic fool of history who tried to build a wall against the ineluctable forces of modernity.

The theme of technical servitude ~~becomes~~ a major point in the macrocosmic social theories of, among others, Thorstein Veblen and Jacques Ellul. In such theories, the presence of modern technics is seen to have both subtle and very obvious shaping effects on the whole range of human behavior, consciousness, and social structure. Seen in their totality, these effects do in fact constitute most of what is important in the life of the individual and in all social relationships whatsoever. Still using the machine model of this state of affairs, Veblen spoke of "the cultural incidence of the machine process,"[53] an incidence he took to be completely overwhelming. "The machine pervades the modern life and dominates it in a mechanical sense. Its dominance is seen in the enforcement of precise mechanical measurements and adjustments and the reduction of all manner of things, purposes and acts, necessities, conveniences, and amenities of life, to standard units."[54] In Veblen's eyes the most important single fact about this state of affairs was that it brought a new and stringent "discipline" to all human activities. A society based on the machine process took on a rigid set of rules, responsibilities, and performance criteria much more demanding of human substance and of social relationships than anything known in previous history. Speaking of the effect on the workman, Veblen notes: "It remains true, of course, . . . that he is the intelligent agent concerned in the process, while the machine, furnace, roadway, or retort are inanimate structures devised by man and subject to the workman's supervision. But the process comprises him and his intelligent motions, and it is by virtue of his necessarily taking an intelligent part in what is going forward that the mechanical process has its chief effect upon him. The process standardizes his supervision and guidance of the machine. Mechanically speaking, the machine is not his to do with it as his fancy may suggest. His place is to take thought of the machine and its work in terms given him by the process that is going forward. His thinking in the premises is reduced to standard units of gauge and grade. If he fails of the precise measure, by more or less, the exigencies of the process check the aberration and drive home the need of conformity."[55]

Veblen argued that the advance of the new technological civilization would displace all previous forms of culture. "The machine discipline," he observed, "acts to disintegrate the institutional heritage, of all degrees of antiquity and authenticity—whether it be the institutions that embody the principles of natural liberty or those that comprise the residue of more archaic principles of conduct still current in civilized life."[56] While there is a lament implied in his observations, Veblen certainly did not wish to stop the movement of an increasingly technologized society. He saw it as an inevitable development, sanctioned by the fact that it was, after all, the true center of the modern condition and an improvement in man's material circumstances. As we have already seen, his criticisms came to rest on the contradictions present in this evolving system, namely, that the persons best suited to operate the machine culture—the engineers—were still subordinate to businessmen.

If one substitutes for the concept of "machine" that of "technique," the position Veblen announced is entirely similar to Ellul's. Both men hold that the technological element has outgrown and absorbed the shell of civilization that once enclosed it.[57] Both men assert that individuals and societies do not rule technical means so much as accept with strict obedience the rule that technical means themselves impose. To describe the technological system, therefore, is to describe the true system of governance under which men live.

Now, it is clear that the condition described by Emerson (the effects of apparatus on the behavior and consciousness of the individual) and Bettelheim (the effects of apparatus on a small group) are not entirely analogous. The macrocosmic social theory is not merely a microcosmic insight writ large. What is similar in these cases, however, is an emphasis upon the context in which tools and instruments operate and have their utility. Such statements ask us to consider what is *required* as well as what is received in the activity of technical "use." Technologies, we noted earlier, are commonly thought to be neutral. The important consideration is how they are used, and this is what permits us to judge them. But, as Ellul points out, the matter of "use" may be entirely settled before one can raise the question at all. "Technique *is* a use," he observes. "There is no difference at all between technique and its use. The individual is faced with an exclusive choice, either to use the technique as it should be used according to the technical rules, or not to use it at all."[58] In this assertion, the broader significance of Emerson's notion that tools are reagents becomes clearer. Complex instruments come equipped with certain rules for their employment, which must be obeyed. People are not at liberty to "use" the instruments in an arbitrary manner but must see to it that the appropriate operating procedures and techniques are followed and that all of the material conditions for operation are met. In modern civilization and its various parts, great amounts of time, energy, and resources are expended in making certain that the procedures are followed and that the conditions are met. Of the meanings of autonomous technology that we have encountered so far, this is the most significant. The technological version of Kantian heteronomy—the governance of human activity by external rules or conditions—is present here as a thoroughgoing yet entirely mundane phenomenon.

There is, of course, a vast multiplicity of such rules and conditions. One might ask, Where are they stated and analyzed in detail? An appropriate place to start would be to examine all of the textbooks in engineering, economics, management, and the various technical skill groups. One obvious but perhaps unavoidable source of incompleteness in the theory of technological politics is that much of the real substance of what it tries to account for is buried in diverse teachings of this sort. It is difficult to footnote or discuss that which is known, practiced, and obeyed in thousands upon thousands of technical specialties.

Indeed, anyone seriously critical of conditions in the technological society soon meets up with the demand from technically trained persons that in order to speak at all, one must first "learn technology." A version of the mode of legitimation through expert knowledge, this advice is, in my experience, usually less a plea for understanding than an urging to compliance. Suggestions for learning of this sort are often made to me at the institute of technology where I teach. They range from the study of calculus, physics, or one of the branches of engineering proper—electrical, civil, mechanical—to a mastery of the techniques of cost-benefit analysis, systems theory, and econometrics. I concede the usefulness of knowing the real activities of such domains of practice and have tried whenever possible to achieve the grasp appropriate to an informed outsider. But given the fact that the specialized fields show important differences in approach and content, the mastery of a particular representative specialty seems only superficially helpful to the effort to comprehend broader situations in which technology is problematic. Those who suggest a technological education of this kind implicitly ask that one undergo a process of socialization. "If you knew what we're doing, then you could not make such criticisms." One comes to appreciate and trust one's professional brotherhood. One comes to accept the virtue of such procedures as quantitive cost-benefit analysis. And, above all, one learns to accept the grand wisdom of the view that the world is a set of "problems" awaiting technically refined "solutions."

In summary, life in a sophisticated technological order supposes that each collection of technical rules, procedures, and trained persons outside one's own sphere of competence must be accepted as given. People are content in the knowledge that things are as they are and that they are in good working condition, and that in society everyone and everything has a certain job to do and does it. And everyone, like every thing, does not find occasion to inquire into this condition or to dispute the manner in which it structures life.

Seen in this light—the ways in which technical rules and preconditions influence human behavior—the traditional notion that technologies are merely neutral tools becomes problematic. Individuals may still retain the noble idea that they can upon sudden inspiration direct the technical means to whatever ends they choose. They tend to see complex technologies as if they were handsaws or egg beaters. Give me a board and I will saw it in half; give me the eggs and flour and I will whip up a chocolate cake. But in highly developed technologies the conditions that make the tool-use notion tenable seldom hold. The technical equivalent of the Archimedean point—a place to put the lever so that one can move the mechanism—is often missing.

Reasons for this state of affairs are apparent in the very nature of modern technologies. Twentieth-century technical devices, as we have described them here, are characterized by enormous size, complex interconnection, and systemic interdependence. In terms of their own internal structure, most of them require precise coordination of the three major elements in our earlier definition. Apparatus almost always requires refined technique: an elaborate, knowledgeable kind of human practice to guarantee its successful working. In the great majority of cases, however, both apparatus and technique require the presence of well-developed, rational, social organization. The world of craftsmanship—the world of technique plus apparatus alone—has vanished. Apparatus, technique, and organization are interdependent, that is, reciprocally necessary for each other's successful operation. This condition has become the sine qua non of all higher technologies of manufacturing, communications, transportation, agriculture, and others. And while there are still small pockets where this kind of interconnected technology is not the rule, such cases are now out of the ordinary.

Such circumstances are of special interest in the theory of technological politics, for two parts of the systemic arrangement—technique and apparatus—require that persons in large numbers be induced to behave according to precise technical principles. Through their "employment" such persons serve a specific function in an organization of many coordinated functions. Since this employment is usually their sole livelihood, there is strong pressure toward strict discipline and obedience. One appears at a preestablished time, for precisely determined work, for an exactly designated reward. This situation is so thoroughly normal in the twentieth century that any sense of how it might be otherwise is largely forgotten. In particular, persons so employed have no sense that the design of the work situation or the character of its operating procedures might be changed through their own conscious intervention. Even more than the employer, technology itself is seen as completely authoritative. It is an *authority* that asks for compliance only and never anything more.

What meaning can the traditional tool-use conception have in this context? Indeed, the structure of men and apparatus is "used" to produce something: goods or services for society. It can also be said that the employees "use" vast technical networks to earn a living. But beyond that, the idea that such networks are merely neutral tools under the control of men, to be used for chosen ends, begins to wear thin. Seen as a way of ordering human activity, the total order of networks is anything but neutral or tool-like. In its centrality to the daily activity and consciousness of the "employee," the function-serving human component, the technical order is more properly thought of as a *way of life*. Whatever else it may be, a way of life is certainly not neutral. Opportunities for "use" or "control" that the human components have within this system are minimal, for what kind of "control" is it that at every step requires strict obedience to technique or the necessities of technical organization? One can say that the "control" is exercised from the center or apex of the system; this is true, although we shall soon see that even this has a paradoxical character. But in terms of the functioning of individual components and the complex social interconnections, "control" in the sense of autonomous individuals directing technical means to predetermined ends has virtually no significance. "Control" and "use" simply do not describe anything about relationships of this kind. The direction of governance flows from the technical conditions to people and their social arrangements, not the other way around. What we find, then, is not a tool waiting passively to be used but a technical ensemble that demands routinized behavior.

In this way of seeing, therefore, the tool-use model is a source of illusions and misleading cues. We do not *use* technologies so much as *live* them. One begins to think differently about tools when one notices that the tools include persons as functioning parts. Highly developed, complex technologies are tools without handles or, at least, with handles of extremely remote access. Yet we continue to talk as if telephone and electric systems were analogous in their employment to a simple hand drill, as if an army were similar to an egg beater.

I am not saying that men and women never use technology or that all ideas of use are nonsense, only that many of our most prevalent conceptions here are primarily nostalgia. There was a golden age when the hand was on the handle and alchemy was the queen of the sciences. But except for the world of small-scale appliances, that time has passed.

The question addressed here can also be posed in terms of the idea of extension. In Emerson's words a full century before Marshall McLuhan, "The human body is the magazine of inventions, the patent-office, where are the models from which every hint was taken. All the tools and engines are only extensions of its limbs and senses."[59] But remembering Emerson's thoughts on tools as reagents we encounter a puzzle. What is an extension of what? Looking at contemporary technologies one sees massive aggregations of human and nonhuman parts, rationally ordered, working in precisely coordinated actions and transactions. Men do indeed claim to be in control and to be the instigators of all the motion. But if one considers the structure and behavior necessary for such systems to exist at all, such claims are either incorrect or ambiguous. When one discovers that people are subtly conditioned by their apparatus, when one learns that their conduct is largely determined by preestablished function and learned technique, when one finds that important social relationships are established according to organizational rationality alone—then the idea of technology as controlled extension becomes entirely misleading. Marx concluded that men had become appendages of the machine in the factory system. In a technical environment that is now more massive, more complete, and more intricate, the conclusion takes on new poignancy.

One can appreciate how it is from this vantage point that automation—the displacement of men and women by machines—diminishes as a problem. The crucial difficulty with the existing technological order is not so much that individuals are "unemployed" by automatic processes (though, certainly, this is a source of grief for a significant minority) but that they are overemployed in ways destructive to their humanity. Marcuse argues that advanced industrial societies do in fact suffer from incomplete and imperfect automation. The productive system still employs human beings as its prime components. "This is," he says, "the pure form of servitude: to exist as an instrument, as a thing."[60] If truly complete automized apparatus were introduced into society, this servitude could be ended for all time. "Automation, once it became *the* process of material production, would revolutionize the whole society. The reification of human labor power, driven to perfection would shatter the reified form by cutting the chain that ties the individual to the machinery—the mechanisms through which his own labor enslaves him."[61] Under the existing system of things, however, automation stands as a threat that the managers of the industrial order hold over their employees as a way of extracting even more toil from them. "At the present stage of advance capitalism," Marcuse writes, "organized labor rightly opposes automation without compensating employment." Labor is thus forced to struggle against its ultimate means of liberation. In this fashion, "The enslavement of man by the instruments of his labor continues in a highly rationalized and vastly efficient and promising form."[62]

Observations and arguments of this kind form the basis of one of the more surprising themes in the literature of technological politics: the myth of labor-saving technology. No one denies that techniques and instruments save time and effort in the performance of specific tasks. And no one denies that the collectivity of such devices enables a society to do things it otherwise could not accomplish or to accomplish them more economically. But one can ask whether the technical innovations added to civilization in the last two hundred years or so have in every case "saved labor" in the sense of lightening man's toil.

Since Marx we have known why this question must be asked in terms of "whose labor?" and "under what conditions in society?" Eli Whitney's cotton gin was a marvelous labor-saving device. Yet its introduction into the system of production in the South actually prolonged slavery and increased the degree of toil extracted from the slaves. The factory system was also a labor-saving innovation, yet there is good evidence that it increased the hours, exertion, and suffering of the workers in the early years of the industrial revolution.[63] Marx's analysis of how such things occur is well known to the reader. What is of interest here is that there exists a distinctive explanation of this phenomenon from the technological perspective. Many have argued that the very nature of advanced technologies—putting aside the matter of ownership and class structure—demands much more of the human being than any previous productive arrangement. The technological order, no matter who *owns* it, is not very efficient at allowing men and women to bank the labor that techniques and instruments have saved. A classic, albeit excessive, statement of this view is found in Friedrich Georg Juenger's *The Failure of Technology*. "Never and nowhere," Juenger argues, "does machine labor reduce the amount of manual labor, however large may be the number of workers tending machines. The machine replaces the worker only where the work can be done in a mechanical fashion. But the burden of which the worker is thus relieved does not vanish at the command of the technical magician. It is merely shifted to areas where work cannot be done mechanically. And, of course, this burden grows apace with the increase of mechanical work."[64]

Juenger's account goes astray in its exclusive emphasis upon the situation of manual labor. Developments in machinery have not had the uniform impact he supposes. Labor-saving devices and automation have eliminated certain tasks, refined others, and created whole new vocational categories. But the broader point that underlies his contention is both valid and significant. Under the relentless pressure of technological processes, the activities of human life in modern society take place at an extremely demanding cadence. Highly productive, fast-moving, intensive, precision systems require highly productive, fast-moving, intense, and precise human participants. The computer has been an especially powerful goad in this direction. Its capacity to do prodigious amounts of work in a very short time puts the humans in the "interface" in a frantic struggle to keep up. The virtues of slow information processing and labor done at a leisurely pace have long since been sacrificed to the norms of work appropriate to the electronic exemplar. The idea that a task is something to be pondered or even savored is entirely foreign to this mode of activity. A telephone call and instantaneous computer check can reserve a room in any of thousands of hotels and motels in a particular network. Since dozens of similar transactions are completed each hour, the employee who does the job cannot spend more than a few moments on any particular request, although certain superficially courteous catchwords may still be part of the rationalized process. What one no longer expects is the innkeeper's handwritten note, received after a characteristic three weeks' delay, which remembers some small detail of your last year's visit.

Pressures of pace are not, however, limited to work environments. They now include the full spectrum of activity involving travel, communication, leisure time, and consumption. Hannah Arendt notes a shift in the focus of human energy that has taken place in the twentieth century. "The two stages through which the ever-recurrent cycle of biological life must pass, the stages of labor and consumption, may change their proportion even to the point where nearly all human 'labor power' is spent in consuming."[65] It is the intensity of the combined activities of labor and leisure that Jules Henry has labeled the "technological drivenness" of modern culture.[66] With Kenneth Keniston, Henry links this phenomenon to the rapidity of technological change. But the connection between pace of work and rate of innovation is by no means a necessary one. Technologies need not be changing rapidly to demand high performance and a rapid tempo of existence.

It is true that technological society is not the first kind of organization to have placed heavy demands upon its members. There is always a price to be paid for culture, for social order, and material well-being. A question raised by the theory of technological politics, however, is whether the price now extracted goes far beyond reasonable limits. How much servitude to technical means is too much? At what point does dependency upon complex systems become a condition of virtual enslavement? The search for criteria upon which one might begin making judgments on these questions offers a rich but as yet relatively poorly explored field of inquiry. Marcuse addresses the issue head on from one direction. He argues that the burdens of civilization as measured by "surplus repression"—"the restrictions necessitated by social domination" above and beyond "the modifications of the instincts necessary for the perpetuation of the human race in civilization"—are now greater than in any previous historical period.[67] The lid upon man's erotic instincts has been screwed down tighter than ever, far beyond any reasonable obedience to the reality principle. Ellul renounces Marcuse's Freudian outlook on the matter.[68] But he agrees that the technological order subjects man to pressures and limitations that are clearly pathological. Not without nostalgia, Ellul describes a condition "common to all civilizations up to the eighteenth century" in which techniques were local and limited, a part of culture rather than its whole.[69] "Man worked as little as possible and was content with a restricted consumption of goods . . . a prevalent attitude, which limits both techniques of production and techniques of consumption."[70]

This state was abandoned when men in Western society perceived the inestimable boon that *la technique* promised.

Whether explicitly stated or strongly implied by those who adopt its vantage point, the theory of technological politics always proceeds with an understanding of limits. Its criticisms point to a boundary beyond which technical artifice no longer enables or liberates mankind. In its evolution, technology arrives at a turning point after which it tends to thwart rather than facilitate the building of an emancipated society. The problem of specifying more clearly what the conditions of human liberation and social emancipation might be is an ambitious project, one that I shall only be able to touch upon briefly in the last chapter. Most of the analysis at this point must attend, perhaps even to a fault, to circumstances of pathology and excess—a corrective to the dewy-eyed traditional assumptions about tools, mastery, and endless benefit.

One is entitled to ask how much of the condition described enters the awareness of persons who live in this world. Do they notice the costs? The answer must be a qualified "yes." In one sense, there is nothing that men and women who live and work in the technological society understand better than the basic conditions which enable this system to function. But their awareness has an intuitive, largely passive quality. The influence of large-scale technical networks is so pervasive and indelible that few of us find occasion to wonder at their effects. We know that "this is how things work." We know that "this is how I do my job." The technological order includes a notion of *citizenship*, which consists in serving one's own function well and not meddling with the mechanism.

In general, then, we live with the costs and do not make the connections as to their origin. Thus, a yawning crevasse opens between the dream of progress and its fulfillment. Men convince themselves, as Ellul points out, that they are about to enter a paradise "in which everything would be at the disposal of everyone, in which men, replaced by the machine, would have only pleasures and play."[71] "In practice, things have not turned out to be so simple. Man is not yet relieved of the brutal fate which pursues him."[72] As a consequence, the citizen of technological society feels a growing frustration and begins the dangerous business of seeking a scapegoat, "the foe who stands in his way and who alone has barred Paradise to him."[73] "He is seized," Ellul continues, "by a sacred delirium when he sees the shining track of a supersonic jet or visualizes the vast granaries stocked for him. He projects this delirium into the myths through which he can control, explain, direct, and justify his actions . . . and his new slavery. The myth of destruction and the myth of action have their roots in this encounter of man with the promise of technique, and in his wonder and admiration."[74] Rather than question the myths of technics, modern man prefers to conjure a malevolent "other" to account for his difficulties.

Transformation and Incorporation

If there is a central hypothesis characteristic of the theory of technological politics, it is this: Once underway, the technological reconstruction of the world tends to continue. The elaboration of rational artifice on a large scale requires that virtually everything in reach be transformed to suit the special needs of the technical ensemble. Anything that cannot be adapted (for whatever reason) is eliminated. This is true of all material parts, all individual human parts, and all segments of the social system. We caught a glimpse of this state of affairs in our discussion of the technological imperative. At that time it was suggested that in a true sense technological structures require the restructuring of their environments. We are now prepared to investigate this contention further to see how technology becomes a force for the total adaptation, integration, and incorporation of the material and human world.

In its quantitative dimension, technological transformation can be seen as a consequence of the scale of modern enterprise. Our vast technical systems have extremely high resource demands. Since more is attempted, more is required. As a result, more of the available world—both material and human—is removed from its original context, defined as a "resource," and brought into a functional position. An important difference between modern technics and those of earlier times is the sheer mass of the necessary means of supply and the size of the area covered in order to obtain these means. Primitive technologies relied upon relatively localized resources. Since the industrial revolution, however, the geographical bases of supply have expanded first to national and then to global levels. It is now physically impossible to maintain many kinds of advanced technologies unless one has a worldwide access to resources. In an early work, Lewis Mumford argued that this fact provided much of the impetus for the international expansion of nineteenth- and twentieth-century "neotechnic" societies. "Under the neotechnic reign their independence and their self-sufficiency are gone. *They must either organize and safeguard and conserve a worldwide basis of supply, or run the risk of going destitute and relapse into a lower and cruder technology.*"[75] Depending on one's point of view, this condition can be seen as an occasion for international cooperation or for an intense variety of imperialism.

The qualitative aspect of the phenomenon is apparent in the manner in which technological artifice, proceeding by division and complex interconnection, modifies whatever it employs. Fabrication now rests upon prefabrication. The making of something means that other things have to have been made and assembled previously. The parts to be connected must be constructed in such a way that they fit in the right place at the right time in precisely the right fashion. The idea of "parts" here includes not only physical objects but also parts in the sense of segmented actions or procedures. Technology takes things that have always existed—physical objects, physical laws, human activities, social relationships—imposes a new form on them, and mobilizes them for a productive purpose. Civilization has, of course, always included implements, regularized ways of doing things, and social organization. But it is only with the coming of modern technology that all such things become subject to the kind of deliberate, coordinated chains of fabrication through prefabrication that we experience today. In one of his characteristic apothegms Ellul observes, "Technique advocates the entire remaking of life and its framework because they have been badly made."[76]

In our accustomed ways of thinking, technologies are evaluated in terms of promise and product. Improvement, usually defined by some highly specific goal, is the subject of concern. The theory of technological politics, however, looks for meaning in questions which have been buried in the process of technical advance. Among the most important of these concern *qualities* sacrificed in technical modification. Things taken apart and reassembled or resynthesized are never the same as before the process took place. Something—functional performance—is gained. But something is also given up. Very often one does not know in advance how great the toll will be.

One way to begin thinking about the sacrifice of qualities is to notice that the success of technology is almost always the victory of artificial complexity over natural complexity. Technological structures and processes are built from complex systems in nature that have been altered in such a way that their substance can be put to use. Alterations of this sort are frequently made with the idea of making things simpler than they had been before. But we have already seen how simplifications on the scale now attempted actually involve very complex linkages. As technology advances, a world of artificial structures replaces the world of complex structures given in nature; thus, people no longer live in anything remotely resembling a natural setting.[77] Technologies are present not only in their specific work but also as a generalized environment for human existence. Taking this point of view, two important questions arise. Which qualities are permanently altered or destroyed by the modification of natural structures? Which new qualities are created, including those neither expected nor desired, by the new structures?

The concept of "nature" here, however, is actually not sufficient for the point to be made. Sometimes one does wish to speak of the modification of things which are in the true sense natural—minerals, metals, plant and animal life, ecosystems, and the other elements and patterns given in nature. The familiar argument of the ecologists is that many natural systems can be modified only at great peril. This is due specifically to the rich complexity of such systems that man does not fully comprehend or, if he does, ought to respect. Tampering with the natural arrangement, the argument goes, often entails costs or risks that may prove to be highly unfortunate.

In other instances, however, the issue has nothing to do with "nature" in its proper meaning at all. The concern is, instead, with the substance of human culture and human character and the ways in which this substance is transformed, either directly or indirectly, by technological modifications. I am referring here to the changes which take place in ordinary language, traditional social institutions, earlier kinds of artifacts, human identity, personality, and conduct through the direct intervention of modern technics. Again, the view often taken in theories of technological politics is that important qualities of human life are sacrificed in this process while new, usually less desirable qualities are substituted.

Consider, for example, the encounter of an individual with his work "role" in an organization. The whole person in its rich complexity of talents, needs, interests, and commitments is of no use in the performance of the role. Instead, only certain selected traits, often created by the role itself and unknown in the life of the individual previously, are demanded by the organization. For much of any given day, therefore, the complex character of the whole person is deliberately contained. Much the same can be said of groups and individuals who encounter bureaucratic techniques. A wide variety of problems, needs, requests, and complaints meet up with the "form" through which they must be communicated. But the efficiency of the form is geared to bureaucratic processes. It does not aim at being faithful to the original feelings and expressions of the persons involved. Thus, much of what was perhaps most crucial in the original situation is simply denied. There is a considerable difference between what human life contains and what technology requires or permits—at first.

One very common but drastically unhelpful way of thinking about the matter centers on the conclusion that certain technical innovations are inherently "dehumanizing," that they violate the essence of man. In most statements of this position the question of what is truly human is never explicitly addressed. If one finds truth in the view present in Marx as well as in modern anthropology that "man makes himself," then the plea to an unspecified human essence becomes flimsy indeed. On what grounds can one say that women and men of the twentieth century are any "less human" than those of the nineteenth century or any other previous era?[78] The concept of dehumanization is at once too specific—calling attention to transformations in humans only—and too vague, begging the question of what "distinctively human" actually is.

But a different formulation can salvage the problem which the abused concept tries to identify. Speaking of human character and culture it is misleading to say that artificial complexity triumphs over "natural" complexity. Neither are we justified in concluding that the technological displaces the purely nontechnological. A better way of expressing the point is this: *More highly developed, rational-artificial structures tend to overwhelm and replace less well-developed forms of life.* In modern history this process has affected virtually every corner of human existence. An entity that was well established in a sphere of practice, thoroughly useful, rich in its complex relationships in the world but not rigorously planned or rationally arranged meets a highly productive technology with a rationalized complexity of its own. If the encounter between the two entities is purposive—that is, if there is a specific end to be accomplished—then the less well-developed form of life, all things being equal, must yield. Its existence and structure may make sense within the broader spectrum of life's activities but fail utterly in the new specified performance required. The structure of the entity—whether it be an implement, a way of speaking, a way of thinking, a mode of behavior, a pattern of work, a social institution, a custom—is interrupted and reconstructed to suit the requirements of the new arrangement of which it is to become a part. Or it is simply replaced. What is cut out in this process, what is lopped off, inadvertently damaged, or permanently excluded may never reach anyone's awareness. Sometimes the awareness may be present, but calculations of cost and benefit accompanying the alteration stipulate that the price is worth it. An urban renewal high-rise is better than an old, decaying section of the city; prepackaged frozen foods are preferable to those prepared at home; programmed learning is superior to the idiosyncrasies of the teacher and classroom. With increasing frequency the analyses and calculations upon which such choices are made are themselves manifestations of a variety of transforming technical artifice, a thoroughly technicized way of handling the formerly nontechnical activity called judgment. Thus, in an important sense, the issue and outcome are predetermined.

One can of course imagine technologies which do not do this—do not cut, shred, rearrange, or lop off significant previously existing qualities. One can even imagine techniques whose main purpose would be to respect or even enhance these qualities. But for the most part, these are not the procedures that we currently find useful.

One of the few scholars ever to tackle questions of this kind on a broad scale was the Swiss historian of architecture, Siegfried Giedion. His *Mechanization Takes Command* was a preliminary investigation of what he hoped would be a thorough-going "*anonymous history* of our period, tracing our mode of life as affected by mechanization."[79] Giedion was convinced that "means have outgrown man," and he sought to demonstrate how this was true in the progress of technical apparatus and procedures, as well as their products. His book is a veritable catalog of the ways in which artificially complex structures confront and modify the structures of nature, society, and human behavior.

Giedion finds the crux of the world-transforming process in mechanical, artificial patterns of matter in motion. For example, Frederick Winslow Taylor's scientific management time and motion studies as applied to work are juxtaposed, with some irony, with a description of "Automatic Hog-Weighing Apparatus for Use in Packing Houses."[80] The mechanization of life means that things are altered so that they can be moved efficiently from one place to another. In this regard he finds the basic principle of all mechanization "in replacing the to-and-fro action of the hand by continuous rotary movement."[81] Giedion traces developments of this kind in a wide variety of examples from agriculture, the industrial assembly line, household appliances, furniture, and other spheres. While most of his analysis is posed in terms of the machine model, he is aware (in 1948) that present and impending technologies will strike off in different directions. There is a new epoch coming, he concludes, "whose trend is away from the mechanical. It centers . . . around man's intervention with organic substance. Animals and plants are to be changed in their structure and in their nature."[82]

The strength of Giedion's discussion is its sensitivity to the subtle alterations that take place in the life of the individual and society through mechanical innovation. He pays considerable attention, for example, to the mechanization of the bath and its role in social intercourse. Bathing, he argues, has been useful historically in two different ways—as the external ablution of separate individuals or as a common social event aimed at total regeneration. "This century, in the time of full mechanization, created the bath-cell, which, with its complex plumbing, enameled tub, and chromium taps, it appended to the bedroom." This arrangement, of course, has no thought of bathing as a social institution or a means of regeneration. "A period like ours, which has allowed itself to become dominated by production, finds no time in its rhythms for institutions of this kind."[83]

Once underway, the technological reconstruction of the world tends to continue without limit. The world moves in transition between two different states of equilibrium. Complete renovation of reality will not cease until the equilibrium of a thoroughly technological world order is established. Ellul argues that at the birth of this process in the eighteenth century, modern technique stood face to face with circumstances totally unsuited for its existence. Traditional culture contained all of the wrong things—the wrong social institutions, the wrong language, the wrong ideas, the wrong tools, and the wrong procedures. The history of the past two centuries amounts to a gradual retailoring of the universe as an appropriate host for the technical phenomenon. "Technique effects its whole operation with completeness," he observes. "It is useless to set limits to it or seek some other mode of procedure."[84] "There is no accommodation with technique. It is rigid in its nature and proceeds directly to its end. It can be accepted or rejected. If it is accepted, subjection to its laws necessarily follows."[85] When Ellul argues this position carefully, when he transcends his reductionist bent and verbal mystification, his case rests on the conflict of complexities of the kind we have just seen.

Ellul's analysis of contemporary techniques of public opinion measurement is a good illustration. 'This system," he observes, "brings into the statistical realm measures of things hitherto unmeasurable." At the same time, the method "effects a separation of what is measurable from what is not." "Whatever cannot be expressed numerically is to be eliminated from the ensemble, either because it eludes numeration or because it is quantitatively negligible."[86] Opinions that can be measured are transformed into aggregate statistical expressions. Those defying such expression are simply not included. For Ellul, this amounts to a "procedure for elimination of aberrant opinions," an unintended side product of opinion methodology. Ellul concludes, 'No activity can embrace the whole complexity of reality except as a given method permits. For this reason, this elimination procedure is found whenever the results of opinion probings are employed in political economy."[87]

Experiences that lend support to the point are now common in everyday life. In a recent encounter I was visited by an opinion researcher who wanted to discover my "attitudes about the gas company." Thinking for a moment I replied, "I'm not sure I have any attitudes about that." "Well," he said, "let me ask you some questions." The man removed a printed form from his briefcase and announced, "How would you respond to the statement: 'Employees of the gas company are prompt and courteous'? Strongly agree, agree, disagree, strongly disagree, or no opinion?" Answering to the meticulously patterned set of queries, my attitudes were in a true sense created right there on the spot. After an item that asked me to rate the gas company's overall service, I explained that I did have some strong views on public utilities in general, mainly having to do with the fact that in the United States they are still privately owned. His questionnaire had no way of recording those opinions, much less the reasons behind them. Yet somewhere, I am certain, my responses, put together with thousands of others, will be counted as adequate "feedback from the public," guiding a better administered firm.

Giedion and Ellul trace the process of transformation and elimination through a herd of examples covering many kinds of activities and products. Both find it significant, for instance, that a simple product like bread had its qualities so radically transformed to accord with the needs of factory production. The multiplication of such illustrations, sometimes bordering on tedium, is an important part of the demonstration; for even though one cannot examine each instance of the transformation described, one can understand that it is a very general process, perhaps even a total one.

Of all the things inappropriate to the man-made environment of the modern age, none is so inappropriate as man himself. "He must adapt himself," Ellul comments, "as though the world were new, to a universe for which he was not created."[88] Some of the adaptation takes place through deliberate manipulation with social-psychological methods, drugs, propaganda, and other "human techniques."[89] Much of it, however, occurs through a quiet, personal adjustment of each individual's behavior, attitudes, ideas, needs, and commitments to the world that surrounds him. Ellul holds that what emerges from these modifications is a human "type" uniquely suited to the technological society—a person who performs a particular work role efficiently, helps improve the technological ensemble whenever possible, willingly receives the pleasures granted, and never raises any serious objections to the conditions under which he or she lives. "It is to be understood . . . ," Ellul adds, "that there is no absolute obligation for the individual to conform to the type. He can, if he will, despise it. But then he will always find himself in an inferior position, vis-a-vis the type, whenever the two come into competition. Our human techniques must therefore result in the complete conditioning of human behavior."[90]

Technical adaptation, then, is in part voluntary, but in some ways it is forced upon the individual. The ultimate means of enforcement is competition among forms of life, a contest which extends throughout the whole of culture, not just the economy. One may, if one wishes, isolate a certain trait or activity for preservation against the influence of technique. But in the end, if that entity plays no role in the technical ensemble, it must either perish or become a mere museum piece. That which is eliminated from the technological order soon appears unreal. That which cannot contribute to the realm of efficient operations is made to seem utterly fantastic.

In its vision of mankind retreating under the technical onslaught, the theory reveals its actual, deeper pessimism, as opposed to the superficial gloom usually attributed to the critique of technological society. Notable pessimists of the past—Diogenes, Seneca, Schopenhauer, Samuel Clemens—thought the human race to be a hopeless, irredeemable species because it could never significantly change. Human nature was seen as fixed and unalterable, always and everywhere the same. Therefore, all attempts to improve mankind were doomed; the indelible human quality with all of its ghastly shortcomings would always win out. The philosophical optimists in the tradition of Western thought have, on the other hand, argued that mankind is capable of positive change. Through proper education or modification of social circumstances, one may expect that a distinctly better human variety will be produced.

The theory of technological politics gives this controversy an odd twist. Ellul, for example, does not side with the pessimists' contention that human nature is debased and unchangeable. On the contrary, he accepts the notion that all of the qualities of man are subject to radical modification and "improvement." But, as Rousseau pointed out two centuries earlier, this trait of perfectibility is precisely the source of man's downfall. Because mankind is totally malleable, it is totally defenseless against any and all attempts to operate on the human character. Ellul's regret comes in noticing that the vaunted improvements are in fact narrowly conceived and made without proper knowledge of or respect for what existed in the original. His eye here is that of a Christian theologian. Through his rendering of psychological, sociological, and political scientific data we can see Ellul's conviction that the rich and marvelous complexity of God's creation, including the human species, is being supplanted by this reckless, rigid complexity of a myriad of harebrained schemes. Technique in his view is truly *sin,* exactly the kind of sin one would expect from a being that had eaten of the tree of knowledge. But the tree of absolute knowledge and the works that spring from it carry the marks of an inevitable ignorance. "For we never know whether there is not something in man which our analyses and scientific apparatus are unable to grasp."[91] Thus, as Karl Jaspers also observed, man lives in a world of self-created imperfections, many of which become an integral part of his character, yet he believes and cannot be convinced otherwise that technology is perfect.[92]

There are ironic passages in Ellul's writing where he notes that knowledge occasionally catches up with technological action just in time to show what is being lost. One such instance is the case of "economic man." Ellul argues that this person was at the outset a purely theoretical construct of nineteenth- and twentieth-century economists. But now, he says, "The human being is changing slowly under the pressure of the economic milieu; he is in process of becoming the uncomplicated being the liberal economist constructed."[93] Man has actually become homo economus, the producing, consuming creature described by the model. All of his traits other than those appropriate to economic-technical performance are now devalued and vanishing. But, as chance would have it, these developments occur "at a time when the theoretical economist is beginning to take account of the real complexity of man, a complexity which, however, man is in the process of losing (if he has not already lost it altogether). The result is that the modern economist still runs the risk of theorizing about an abstraction because he is speaking of a man philosophically conceived or of some historical or traditional image."[94] Like Orpheus turning to gaze at Eurydice, what we see is a faint image rapidly receding into nothingness.

It is important to add that an awareness of the transforming, incorporating aspects of technological refinement is not the sole property of the critics of technological society. Among certain technical professions it has become an item of pride that the building of a new system entails complete retailoring of the forms of life that the system touches. In his study of computers and management, Michael Rose explains: "The most important of specialisms are those of the people who reduce the complexities of a commercial or manufacturing process to an explicit routine, who transform such routines to sets of detailed instructions to the machine, or who supervise the work of those engaged on these tasks and take general responsibility for advancing the progress of a computerization project."[95] "A proficient systems man," he notes, "has always been as much a designer as an analyst, not merely establishing a model of the system as it exists but reshaping it into a new and more effective, computer-oriented information flow."[96] Indeed, theory we are considering could well be pieced together from ideas and observations of the engineers themselves rather than from the writings of philosophers. The point is, however, that engineers themselves seem unwilling to ponder their work from any general political or philosophical **perspective and even seem to resent the fact that Ellul, Mumford,** Marcuse, and others have taken up the task.[97]

It goes without saying that ideas of the kind we are examining present a rich field for the intellectual historian. They bear a definite resemblance to the belief of many poets and playwrights of the early nineteenth century that the essential wholeness of the world faced imminent destruction at the hands of liberal, bourgeois, rational, utilitarian society.[98] Views of this sort were especially prevalent among German writers of that period and later experienced a revival in the drama, poetry, philosophy, and social criticism of the Weimar Republic.[99] Several of the writers one could list as theorists of technological politics—Juenger, Heidegger, and Jaspers, for example—were clearly influenced by the earlier, anti-Enlightenment thought of Friedrich Hölderlin and others. Twentieth-century critics of technological society are not without precursors.

But in the work at hand, I have chosen not to engage in the search for distant roots or lines of intellectual influence. Someday, I expect, this will be done in a different sort of book. But since I want above all to treat these themes as problems relevant to our immediate situation, I must leave the details of an intellectual history to someone with a different project.

Our treatment of technological transformation still leaves one significant aspect of the phenomenon untouched: its mode of enforcement. The theory holds that once something is restructured and incorporated into the technical ensemble, it becomes an active part of the defense of the entire system. Neither the technological order nor any of its significant parts can long tolerate opposition, defective performance, or idiosyncrasy. Such characteristics are dysfunctional; left unchecked they tend to breed the disorder of apraxia within the system of interdependent parts. Protection against them is sometimes enacted through force—the objectionable part of activity is rooted out and done away with. But a more common and more important response involves no use of overt force at all. In the normal working of things, sources of opposition or idiosyncrasy are simply isolated, neutralized, and then reincorporated as functional parts of the order itself.

An account of this process lies at the heart of Herbert Marcuse's discussion of one-dimensional man in technological society. The course of modern history, he argues, has produced a social system, a language, a way of knowing, and a human temperament that permit no lasting opposition. The dialectical forces in historical thought and social existence Hegel and Marx described have been modified by the unrelenting process of modern technological transformation. Where one formerly would have found the creative clash of opposites, one now encounters a repetitive reaffirmation of efficient, productive technical solutions in a superficially pleasant world. The multidimensional structures of man and society in previous history—pregnant with revolutionary possibilities—have been supplanted by one-dimensional structures appropriate to standardized, functional designs. Most notable of the entities neutralized by this condition is none other than the proletariat, whose opposition, according to the Marxist vision, was counted upon to turn the technological nightmare of capitalism into the technological utopia of communist society. In Marcuse's view, "The new technological work-world . . . enforces a weakening of the negative position of the working class: the latter no longer appears to be the living contradiction to the established society."[100] But even more significant than this is the fact that the technological universe has absorbed and directed to its own purposes every conceivable variety of opposition, not merely those of proletarian origin. Mentioning the alternatives supposedly offered by metaphysics, spiritual occupations, existentialism and bohemian life styles, Marcuse comments, "Such modes of protest and transcendence are no longer contradictory to the status quo and no longer negative. They are rather the ceremonial part of practical behaviorism, its harmless negation, and are quickly digested by the status quo as part of its healthy diet."[101] Co-opted, popularized, and made socially acceptable, deviance is a useful safety valve for a society, one of whose remaining problems is psychological tension.

As the means of technological domination have become more extensive, so have they grown more subtle. For Marcuse, as for Ellul, the enforcement of the new world order is carried very lightly within each individual and within every activity of human culture. There is no need for terror or violence to bring about compliance. The adaptation of men has been thorough. Each person simply obeys the performance criteria appropriate to his station and happily receives the promised rewards: security, leisure, and material goods. The technological society is one in which obvious social "needs" are fully taken care of. Even the desire for freedom is preserved in the arena of consumer preference. Shopping centers become the public space for the exercise of human liberty. Through a wonderful coincidence (which we shall examine later) the free choices produce an aggregate "demand," which matches exactly what the system of production is best able to provide. The only political questions that remain are those of distribution. Marcuse and Ellul note that there are still important differences in the distribution of wealth and privilege within the system; but these do not cause major conflicts—certainly no revolutionary disorders—since most people are sufficiently satisfied in their needs to ignore inequalities of class. Given a large enough slice, men do not worry much about comparisons or about the origins or nature of the pie.

But even if one did have a mind to raise political questions of a different sort—questions about the structure of this order or its desirability as a way of life—it would be virtually impossible, for the very language that one could use to think about and communicate these concerns is itself a one-dimensional medium, which subtly undermines any attempt to raise serious questions in opposition. Marcuse is especially emphatic on this point. He finds in the styles of discourse of positivism, behaviorism, functionalism, operationalism, orthodox social science, and advertising a uniform tendency to turn everything thought or spoken into a fruitful affirmation of the technological universe. All thinking of a different inclination is subverted or jammed by a harmonious structure of positive language, which insists that a thing can be described in terms of its function. "This language controls by reducing the linguistic forms and symbols of reflection, abstraction, development, contradiction; by substituting images for concepts. It denies or absorbs the transcendent vocabulary; it does not search for but establishes and imposes truth and falsehood."[102] "Functional communication is only the outer layer of the one-dimensional universe in which man is trained to forget—to translate the negative into the positive so that he can continue to function, reduced but fit and reasonably well."[103]

Unfortunately, in making his point Marcuse overshoots the mark, for he includes among the proponents of one-dimensional thinking British and American ordinary-language philosophers, one of whose primary accomplishments has been to demonstrate the rich multidimensionality of natural language and to employ it as a key to the investigation of philosophical questions. Marcuse's opinion appears to be that the only exceptions to the homogenization of speech are to be found among those who employ the linguistic fashions of Hegel, Marx, and the Frankfurt Institute for Social Research. The interesting modes of criticism found in Wittgenstein, J. L. Austin, and other ordinary-language philosophers evidently do not count.

The argument is more appropriately aimed at the widespread influence of systems theory upon contemporary thought. Here the leveling, neutralizing tendency that Marcuse describes is strongly present. Central to the various manifestations of the systems approach is the attempt to describe with total economy the structure and performance of self-regulating systems whose telos is a state of equilibrium. Sources of conflict, disruption, or opposition are defined as dysfunctional states, aberrations to be overcome as soon as possible. Systems analysis, furthermore, tends to disregard the complexities of ordinary language in talking about human situations and social institutions. A substitute set of terms borrowed wholesale from modern engineering—input, output, hardware, software, interface, feedback, programming—is considered adequate to express anything of importance. Combined with the block diagram, flow chart model of the universe and its parts, this approach amounts to a singleminded vision of reality, perhaps *the* modern vision. "There seems to be no substitute for or no equivalent to the block diagram," remarks one of the enthusiasts of this school, "as a technique for defining and communicating what is required at all levels, from broad objectives to detailed physical solutions."[104]

Those who now unconsciously employ such terms as "input" and "feedback" to refer to human communication forget the origins of such words and the baggage in meaning and sensibility that they carry. Feedback, some will be surprised to discover, is not the same as a response. Over the past two decades the language and mentality of systems analysis have become an increasingly central part of political awareness in advanced industrial nations.[105] A comparison of the modes of expression in public discussions of our own time to those of the 1930s reveals the direction of the drift. In the earlier period political language included a substantial emphasis upon the moral concerns of a community suffering certain difficulties. Looking at issues of unemployment, hunger, old age, and human welfare, public spokesmen were apt to raise questions about the "responsibilities" or "obligations" of the society to persons in need. Today the tendency is to see such problems as those of the malfunctioning of a complex social mechanism requiring new "incentive structures" and other "policy tools" to bring the system back to proper order.[106] The traditional language of morality and politics is employed less and less in such accounts and is even seen as an embarrassment. Instead, an educated person is expected to talk and write in the hollow, pseudo-precision of a bizarre tongue which reflects the life of pure instrumentality, to wit: "Estimates of income allocation parameters in lower socio-economic strata indicate trade-off options marked by budgetary constraints."[107]

It is important to notice that the problem we are considering here has nothing to do with the traditional notion of "use" and "misuse." Technological transformation occurs prior to any "use," good or ill, and takes place as a consequence of the construction and operating design of technological systems. The phenomenon is found where an instrument is taking shape as an instrument but before the time when the instrument is employed to do anything. Technological transformation, whether by deliberate action or unconscious adaptation, is an essential part of any modern technology's preparation for performance. This does not mean that the instrument cannot be judged as to whether it is used well or poorly for good or for evil. It does suggest, however, that by the time the issue of "use" comes up for consideration at all, many of the most interesting questions involved in how technologies are constituted and how they affect what we do are settled or submerged. The question: "Was it used well or badly?" is in this sense like asking, "Who's at the bat?" long after the game is over and the score recorded. For this reason, it becomes of paramount importance to examine the structure of technologies, as well as the goals of their employment.[108]

In focusing upon matters of this sort, our theory faces the objection that it is merely antitechnical and antitechnician. Its position, some would maintain, amounts to nothing more than the idea that technology in all of its forms is bad and ought to be avoided.[109] Those who make this charge are convinced that all technologies are merely neutral tools and that the only valid question that an intelligent, honorable person could debate is that of wise and unwise use. But if matters of structure and mode of operation are to be excluded from scrutiny by all but the appropriate experts, then the most crucial aspects of the formative influence of technology in the world are totally removed from any conscious, public attention or dispute. It is in this formative character that technology gives up its claim to neutrality and becomes a distinctly political matter. While it is widely admitted that the structures and processes of technology now constitute an important part of the human world, the request that this be opened up for political discussion is still somehow seen as an attempt to foul the nest.

Here also we see a basic difference that divides the perspective of technological change discussed earlier from the perspective we are now developing. The difference comes in noticing *when* it makes sense to ask a question. Thinkers who focus upon the notion of technological change find opportunities for making judgments and taking action at only those points in which a *new* development in technology occurs. Attention is given to the use and possible misuse of a new invention and to the side effects that a technical innovation might generate. Thus, social scientists are intensely concerned with the economic, social, and political impact of such new technologies as SSTs, VSTOLs, cable television, automated education, satellite communications, and digital computers. These are jazzy topics. They are worth studying and studies on them can usually find funding. But for scholars supposedly worried about the representativeness of their samples, these selections present a remarkably skewed array. They are all technologies at the frontiers of progress, the next novelties around the corner. Never raised for serious consideration are techniques and devices whose development and impact came decades ago and are now part of the structure of the human world order. These are understood to be "given" and unquestionable, not subject to social scientific probing or political dispute.

The theory of technological politics, on the other hand, insists that the *entire structure* of the technological order be the subject of critical inquiry. It is only minimally interested in the questions of "use" and "misuse," finding in such notions an attempt to obfuscate technology's systematic (rather than incidental) effects on the world at large. Similarly, it gives little weight to problems of unintended, undesirable remote side effects, hearing in such complaints little more than a justification for one more layer of technique unenlightened by the slightest awareness of what the world of technique in its totality is about. No; the theory of technological politics is not concerned with alternative uses or side effects. Its direct, sustained, and sometimes annoying fascination is with characteristics that are built into the technological order, aspects of the way in which that order has been constructed. The argument is not that technology is *misused,* but that in a fundamental sense it is *badly made.* The kinds of apparatus, technique, and organization that have been built during the last two centuries are seen to be utterly destructive of much that is good in nature, man, and society, lethal to many positive possiblities.

Our discussion of technological transformation in complex structures tries to clarify the grounds for this variety of criticism. To accept its logic is, indeed, to find oneself opposed to many of the developments commonly trumpeted as "technological progress." But such objections are in no sense blind ones. They amount to a reasoned way of standing "progress" on its head in order to appreciate what is at stake. In a discussion that until recently thrived on unexamined homilies of "growth," "efficiency," "improvement," and "use," it makes sense to reintroduce critical content to the question, What is involved in having a technology?

Reverse Adaptation

The theme of technological transformation skirts an issue which is among the most interesting our perspective discloses. The problem raised in this manner of looking at things is not merely that modern technics gives a new order and discipline to the world. Neither is it enough to indicate that whole categories of structures in nature and human life are qualitatively altered in the process. Beyond this lie cases in which technical systems, once built and operating, do not respond positively to human guidance. The goals, purposes, needs, and decisions that are supposed to determine what technologies do are in important instances no longer the true source of their direction. Technical systems become severed from the ends originally set for them and, in effect, reprogram themselves and their environments to suit the special conditions of their own operation. The artificial slave gradually subverts the rule of its master.

To avoid confusion, we should notice what is not asserted in attempts to describe this phenomenon. No one has written that technical means are totally divorced from human wants and needs, cultural standards, political decisions, individual choices, and the like. No one has argued that technology and human motives have parted company in an absolute way. Ellul, whose positions mark a certain extreme in this regard, still insists that the technological society is based upon the ideals and motives of bourgeois culture (rationality, profit, material comfort, convenience), which now stand as latent, ossified, seldom examined underpinnings of every social practice.[110] He explicitly argues that technique can advance only by satisfying social needs but maintains that the relationship of technique and need amounts to a selective process that eventually imposes a one-sided, tyrannical pattern on the development of human capacities. He takes care to emphasize the fact that human beings are present—desiring, thinking, deciding, acting—at each step in the technological progression. The weight of his message is, however, that such desire, thought, decision, and action are very thoroughly corrupted by circumstances which arise from modern man's adaptation to technique. It is important to mention such things, for the crisis of steering in the technological order does not mean, as some seem to think, that there is no one at the wheel and that the car literally drives itself. It does mean that the relationship between car and driver, continuing the metaphor, is problematic and sometimes not that which ordinary tool-use conceptions lead us to expect.

Under the tool-use idea we tend to believe that there is an entirely obvious connection between the thing desired and the means to its fulfillment. One begins with a preconceived end in mind. Then one decides upon an appropriate instrument or organization of instruments to achieve that end, usually weighing the advantages of two or more alternative methods. Next comes the actual *use* of the instrument in the way established for its successful exercise. Finally, one achieves certain results which are judged according to the original end.

Now, if one thinks about the matter for very long, one notices that in actual practice the straight-line notion of means and ends is often not realized. It is in fact the exception rather than the rule. This idea describes one possible way—the way certified by our cultural ideal—in which ends can be related to technical means. But it is certainly not the only way. If one insisted upon finding the linear relationship in every instance in which persons engage in technical activity, one would be woefully disappointed. Under the conditions characteristic of an advanced technological order—complex interconnection, technical rationality and the vast scale, concentration and interdependence of major enterprises—the old-fashioned, tool-use mode is replaced by many other sorts of relationships between persons, their ends, and the means available.

In the complex, large-scale systems that characterize our time, it is seldom the case that any single individual or group has access to a technological process along the whole of its conception, operation, and result. More common is a situation in which persons have the opportunity to enter into the process at one point only. The most common of roles in this regard is that of the consumer who enjoys the end products of the technology. But if we take *use* to refer to the whole line or sequence of thought, action, and fulfillment, then the consumer's role could not be called *use,* since his appearance comes only at the last stage. Perhaps a better term to describe this situation would be *utilization*—the acceptance and enjoyment of the products of a technological process. *Utilization,* as opposed to *use* in this sense, is a largely passive mode of behavior. It does not ask for participation in the establishment of the goals of the technical system, does not enter into the design of the system or in the choice among alternatives, and is not part of the actual working of the instrument.[111] This applies, by and large, to the relationship most persons have to the technical systems that provide them with the necessities and amenities of life.

Another point at which individuals become engaged along the means-ends spectrum is at the stage of instrumentality—the stage at which the interconnected components of a technical system, including the human parts, must operate. It is here that we encounter important instances of a phenomenon I wish to call *reverse adaptation*—the adjustment of human ends to match the character of the available means. We have already seen arguments to the effect that persons adapt themselves to the order, discipline, and pace of the organizations in which they work. But even more significant is the state of affairs in which people come to accept the norms and standards of technical processes as central to their lives as a whole. A subtle but comprehensive alteration takes place in the form and substance of their thinking and motivation. Efficiency, speed, precise measurement, rationality, productivity, and technical improvement become ends in themselves applied obsessively to areas of life in which they would previously have been rejected as inappropriate. Efficiency—the quest for maximum output per unit input—is, no one would question, of paramount importance in technical systems. But now efficiency takes on a more general value and becomes a universal maxim for all intelligent conduct. Is the most product being obtained for the resources and effort expended? The question is no longer applied solely to such things as assembly-line production. It becomes equally applicable to matters of pleasure, leisure, learning, every instance of human communication, and every kind of activity, whatever its ostensive purpose. Similarly, speed—the rate of performance and swiftness of motion—makes sense as an instrumental value in certain kinds of technological operation. But now speed is taken to be an admirable characteristic in and of itself. The faster is the superior, whatever it may be.

In the vogue for "reading dynamics," for example, we see one instance in which the combined emphasis upon speed and efficiency—increased words per minute plus increased comprehension—reduces an activity with many possible values to a pure instrumentality. Slow reading, for example, can often be a marvelous occasion for thoughtful reflection. Nietzsche believed this and advised that readers take a walk to ponder the things that they found insightful in books. But Nietzsche was not Evelyn Wood. He did not know that the primary end of reading is not insight or illumination but merely the maximum information crammed in per minute expended.

The reader may now want to pause a minute to consider other instances in which things have become senselessly or inappropriately efficient, speedy, rationalized, measured, or technically refined. To ponder so will help make clear what is meant by the primacy of instrumental values. It will, however, cut your reading speed on this section considerably.

The predominance of instrumental norms can be seen as a spillover or exaggeration of the development of technical means. It is not that such norms are perverse in themselves but rather that they have escaped their accustomed sphere. Instrumental virtuosity—shown in efficiency, speed, accuracy, and productivity—is required of individuals in the task performance of their work. But the range of things covered by standards appropriate to instrumental task performance tends to increase. Less and less of the business of living is left to chance. Therefore, as individuals become acclimated to the rule of instrumentality in a central portion of their activity, they extend this rule to everything else as well. Internal consciousness begins to mirror the conditions of the external order. There is no need to say that this is a necessary eventuality. Heroic efforts might well be made to maintain the split between the standards and concerns of our technological environment and those appropriate to the rest of life. Suffice it to say that it is not surprising that such changes do occur and that one is hard pressed to think of how, in the long run, it could be otherwise.

The literature on this subject finds the dominance of instrumental norms and motives enforced by basically two kinds of mechanisms. The first is a psychological formation in which the technically adapted side of one's personality begins to exercise control over the rest of the personality. The second is a condition in social situations such that all problems are ultimately defined in terms of instrumentality and only instrumental concerns have any influence. In both of these categories, the totalism of technological rule becomes more than evident.

Illustrations of the way in which the first mechanism operates are found in the work of Kenneth Keniston, one of the more eloquent and intelligent of the contemporary writers who have taken up the technological perspective as a focus for empirical research. A rarity among social scientists, Keniston finds it possible to investigate technological and social change while at the same time paying attention to political and psychological domination. In the normal practice, one must choose one or the other of these problem sets, a choice that determines where one publishes, which conferences one attends, and who finds one's work "interesting." In his studies of alienated youth, Keniston found that only a portion of the psychological problems he observed could be attributed to the identity crises and "generation gap" associated with technological change. Another part of the syndrome of alienation comes from the influence of "the technological ego." "Probably no other society in human history," he writes, "has demanded such high levels of specialized ego functioning as does ours. Our ability to tolerate those who lack minimal ego competence has decreased; our effort to teach and promote technological skills in family and school has become more and more systematic; the demand for peak ego performance has grown."[112] Keniston holds that by exceeding any reasonable limit, this social-psychological formation has had very destructive effects. In explaining this state of affairs, Keniston employs what is by now a familiar political metaphor. "The virtues of our technological society require a dictatorship of the ego rather than good government. The self-denying potential of the ego is minimized: playfulness, fantasy, relaxation, creativity, feeling, and synthesis take second place to problem-solving, cognitive control, work, measurement, rationality, and analysis. The technological ego rarely relaxes its control over the rest of the psyche, rarely subordinates itself to other psychic interests or functions. Though its tyranny is seldom obvious, it is firm and unrelenting. Although apparently benevolent and reasoning, seeking to 'understand' the motivations it regulates, ignoring the pangs of conscience when it can (and when it cannot, seeking to undermine their claims), the technological ego still dominates rather than governs well."[113]

Similar psychological notions appear in Ellul and Marcuse and are crucial to that part of their arguments concerning how human compliance with the structures of a technological order is maintained. Marcuse's theories of surplus repression and one-dimensionality, for example, come together to form a hypothesis very similar to that which Keniston advances. The important point, and certainly a point that adds to the atmosphere of gloom surrounding such theories, is that the ultimate enforcement of the vast sociotechnical structure's requirements comes from internal (yet entirely artificial) sources. The policing mechanism is none other than the combined force of the individual's own drives, standards, and conceptions of what is desirable. With his gift for a dramatic phrase, Ellul comments: "The new man being created before our very eyes, correctly tailored to enter into the artificial paradise, the detailed and necessary product of the means which he ordains for himself—*that man is I.*"[114] Insight guarantees no special extrication.

There is another important way in which the dominance of instrumental values is insured. It reinforces the mechanism just discussed but rests on an entirely different foundation, for beyond the fact that people experience a psychological obsession with instrumentality, the technological society tends to arrange all situations of choice, judgment, or decision in such a way that only instrumental concerns have any true impact. In these situations questions of "how" tend to overpower and retailor questions of "why" so that the two matters become, for all practical purposes, indistinguishable.

We have already noted that in Ellul's theory important social ends, ideals, and commitments stand at the historical base of technological culture. But all of these have receded into the background and are no longer, he asserts, part of the living process of choice. The ends have become "abstract" and "implicit." They "are no longer questioned."[115] "It is true that we still talk about 'happiness' or 'liberty' or 'justice,' but people no longer have any idea of the content of the phrases, nor of the conditions they require, and these empty phrases are only used in order to take measures which have no relation to these illusions. These ends, which have become implicit in the mind of man, and in his thought, no longer have any formative power: they are no longer creative."[116]

What causes are responsible for this state of affairs? Ellul argues that the withdrawal of the ends of action into an inert, moribund condition comes at exactly the time when the means of action have become supremely effective. The tendency of all people is to hold the ends constant or to assume that they are well "known" and then to seek the best available techniques to achieve them. There is, then, a twofold movement affecting all social practices and institutions: (1) the process of articulating and criticizing the matter of ends slips into oblivion, and (2) the business of discovering effective means and the ways of judging these means in their performance assumes a paramount importance. Thus, new kinds of apparatus, organization, and technique become the real focus for many important social choices. Instrumental standards appropriate to the evaluation of technological operations—norms of efficiency above all others—determine the form and content of such choices. Locked into an attachment to instruments and instrumentalities, social institutions gradually lose the ability to consider their fundamental commitments.

But while ends have become passive in the face of technical means, the reverse is not true. Indeed, the nature of the means requires that the ends be precisely redefined in a way that suits the available technique.

Abstract general ends—health, safety, comfort, nutrition, shelter, mobility, happiness, and so forth—become highly instrument-specific. The desire to move about becomes the desire to possess an automobile; the need to communicate becomes the necessity of having telephone service; the need to eat becomes a need for a refrigerator, stove, and convenient supermarket. Implied here also is the requirement that the *whole chain of techniques and instruments* which satisfies each need is well constructed and maintained. Technique, apparatus, and organization are themselves interdependent parts of the technological structure. The desire for the products of the network amounts to a desire for the continued support of each link in the hookup. Thus, a desire to communicate—an abstract, implicit, seldom-examined end—becomes the active need to support and extend, for example, a nationwide telephone system. The desire for geographical mobility becomes the practical need to build and keep in adequate functioning order a variety of transportation systems—railroads, airlines, bus lines, freeways—in all of their complex connections. Once individual and social ends have become so identified, there is no avoiding this kind of affirmation. I vote "for" Consolidate Edison's full range of technical interconnections every time I switch on my electric typewriter.

Going a step further, there is another distinctive way in which the presence of technique refashions ends. In addition to the techniques immediately and clearly effective in application to practical tasks, there are parallel techniques that measure how well the means are being employed. Techniques of measurement are themselves highly specific. Inherent in their construction is the ability to recognize certain factors while blocking out all others. One asks the question, How well are the technical means doing? The way of answering the query is itself reconstructed to fit the techniques of measurement chosen. Testing, sampling, surveying, statistical evaluation bring together new combinations of physical instruments, methods, and productive organization. Far from being neutral, uninvolved sensing devices, these technical ensembles have their own requirements that must be met if the measurement is to take place. Individuals and social institutions must adapt to these requirements or they cannot be adequately evaluated.

The influence of standardized, centralized, computer-scored mass testing in education is a good example of how this works. The tests can measure only those qualities of a student's education that can be represented in pencil-marked squares on the test sheet during a four-hour examination. On top of this, students and teachers soon learn the game and its stakes. It is not unusual for high-school seniors in the United States to spend the better part of their time mastering the specific kinds of performance likely to appear on College Entrance Examination Board tests. This has progressed so far that there is now genuine concern that elementary writing skills are rapidly deteriorating simply because multiple-choice, computer-scored exams do not measure them. "What the teachers are saying to us," one official of the testing organization remarked recently, "is that if the College Board does not require writing as part of its basic testing program, then writing won't be valued in the schools, and teachers won't require it of their students."[117] Thus, as a result of the structure of the instrument and human adaptation to it, techniques of measurement become purely self-fulfilling.

In one of the most unsettling arguments in his book Ellul maintains that the combined result of developments of this kind is to produce a virtual *automatism* in matters that are ostensibly free choice. The original ends have atrophied; society has accepted the power of technique in all areas of life; social decisions are now based upon the validity of instrumental modes of evaluation; the ends are restructured to suit the requirements of techniques of performance and of measurement. Thus, selections as to *what* is to be done and *how* proceed almost as if by clockwork. Once such conditions take effect, only an extraordinary act of will can reopen the process of evaluation, choice, and action, for all situations in social life turn out to be those in which a known instrument is available to do the job and a corresponding instrument ready to make the right selection. Ellul writes:

> Technique itself, *ipso facto* and without indulgence or possible discussions, selects among the means to be employed. The human being is no longer in any sense the agent of choice. Let no one say that man is the agent of technical progress . . . and that it is he who chooses among possible techniques. In reality, he neither is nor does anything of the sort. He is a device for recording effects and results obtained by various techniques. He does not make a choice of complex and, in some way, human motives. He can decide only in favor of the technique that gives the maximum efficiency. But this is not choice. A machine could effect the same operation. Man still appears to be choosing when he abandons a given method that has proved excellent from some point of view. But his action comes solely from the fact that he has thoroughly analyzed the results and determined that from another point of view the method in question is less efficient.[118]

Thus, by its systematic confounding of processes of thought, motivation, and choice, modern technology tends to remove its workings from effective direction by human agency. The results of this tendency so closely approximate a self-generating, self-sustaining technical evolution that efforts to argue for the reality of human guidance seem completely vain. Of course, *in principle* man is always at the control panel. But this is increasingly a principle hollow of any living substance.

In summary, the position I have described identifies certain processes of selectivity at work in the formation of modern culture, processes generated by the structures and operations of technics itself. My aim has been to give a general account of such phenomena and explain their underlying rationale. This understanding of things does not necessarily exclude all other varieties of cultural selection. There are important factors which act *upon* technics as well. Jacob Schmookler has shown exactly what one would expect—that the market mechanism has had a substantial influence upon the kinds of inventions and innovations developed over many decades.[119] Nowhere does the present theory deny the presence of such economic forces. It merely suggests that such factors may not be the overwhelmingly decisive ones. There is a sphere of vital concern that one misses if every question is quickly reduced to categories of economics.

Chapter 6
Technological Politics

The last two chapters have examined two drastically different ways in which technology can be seen as a political phenomenon. First was the orthodox notion of technocracy, which finds the importance of the matter in the ascendance of a new group of knowledgeable persons destined to gain power in an age of advanced technics. From this vantage point the whole subject appears as little more than a special problem for elite or ruling-class theory. One traces down the identity, social base, and circumstances of influence of the technically proficient group and comes, thereby, to the locus of technology's political substance.

A second approach takes a much broader and more extraordinary route. Here one locates the political essence of technology in its total formative impact on *all* of nature and human culture. Technological politics, in this manner of seeing, encompasses the whole of technology's capacity to transform, order, and adapt animate and inanimate objects to accord with the purely technical structures and processes. It is the system of order and governance appropriate to a universe made artificial. To the extent that the human world becomes a product of rational artifice, it will fall under this mode of governance. Political reality becomes a set of institutions and practices shaped by the domination of technical requirements. The order which evolves is marked by stringent norms of performance, rigid structural limitations, and a tendency to alter subtly the human master's relationship to the technological slave.

This second notion has an unfamiliar ring since it stretches the meaning of the word *politics* beyond its ordinary context. Indeed, the thoughts advanced in the previous section are perhaps best called a theory of culture, a theory of the patterns human institutional life, structures of consciousness and conduct, take in advanced technological societies. It is, nonetheless, a theory of culture with teeth. My selection of the term *technological politics* is meant to emphasize a point made again and again in the sources mentioned—that the rule of technological circumstances in the modern era does in fact supplant other ways of building, maintaining, choosing, acting, and enforcing, which are more commonly considered political. In this chapter I shall further clarify the political edge of the theory. In so doing we will be able to return to and reassess the issue of technocracy. The "who governs" and "what governs" will be at least partially reconciled.

Reverse Adaptation and Control

We have already considered one of the ways in which a series of transformations engendered by the introduction of technological means eventually leads to transformation of ends. I called this process *reverse adaptation,* for the interesting situation is one in which the exact opposite of the idealized relationship of means and ends is the one that occurs. Ends are adapted to suit the means available.

In a somewhat different form, the process of reverse adaptation is the key to the critical interpretation of how ends are developed for large-scale systems and for the activities of the technological society as a whole. Here the conception of autonomous technology as the rule of a self-generating, self-perpetuating, self-programming mechanism achieves its sharpest definition. The basic hypothesis is this: *that beyond a certain level of technological development, the rule of freely articulated, strongly asserted purposes is a luxury that can no longer be permitted.* I want now to state the logic of this position in the language introduced in the previous chapter.

Of interest to the theory are technological systems or networks of a highly advanced development—systems characterized by large size, concentration, extension, and the complex interconnection of a great number of artificial and human parts. Such conditions of size and interconnectivity mark a new "stage" in the history of technical means. Components that were developed and operated separately are now linked together to form organized wholes. The resulting networks represent a quantum jump over the power and performance capabilities of smaller, more segmental systems. In this regard, the genius of the twentieth century consists in the final connecting of technological elements taken from centuries of discovery and invention.

Characteristic also of this new stage of development is the interdependence of the major functioning components. Services supplied by one part are crucial to the successful working of other parts and to the system as a whole. This situation has both an internal and external dimension. Within the boundaries of any specific system, the mutual dependencies are tightly arranged and controlled. But internally well-integrated systems are also in many cases dependent upon each other. Through relationships of varying degrees of certainty and solidity, the systems establish meta-networks, which supply "inputs" or receive "outputs" according to the purposes at hand. One need only consider the relationships among the major functional components—systems of manufacturing, energy, communications, food supply, transportation—to see the pulse beat of the technological society.

Large-scale systems can succeed in their ambitious range of activities only through an extension of *control.* Interdependence is a productive relationship only when accompanied by the ability to guarantee its outcome. But if a system must depend on elements it does not control, it faces a continuing uncertainty and the prospect of disruption. For this reason, highly organized technologies of the modern age have a tendency to enlarge their boundaries so that variables which were previously external become working parts of the system's internal structure.

The name usually given to the process of thought and action that leads to the extension of control is *planning,* which means much more than the sort of planning done by individuals in everyday life. Planning in this context is a formalized technique designed to make new connections with a high degree of certainty and manipulability. Clear intention, foresight, and calculation combine with the best available means of action. In some typical passages from Ellul we read:

> The more complex manufacturing operations become, the more necessary it is to take adequate precautions and to use foresight. It is not possible to launch modern industrial processes lightly. They involve too much capital, labor, and social and political modifications. Detailed forecasting is necessary.[1]

> Planning permits us to do more quickly and more completely whatever appears desirable. Planning in modern society is *the* technical method.[2]

In the complexity of economic phenomena arising from techniques, how could one justify refusal to employ a trenchant weapon that simplifies and resolves all contradictions, orders incoherences, and rationalizes the excesses of production and consumption?[3]

Size, complexity, and costliness in technological systems combine to make planning—intelligent anticipation plus control—a virtual necessity. This is more than just convenience. Planning is crucial to the coherence of the technological order at a particular stage in its development.

One can ask, What would be the consequences of an inability to plan or to control the span of interdependencies? A reasonable answer would be that many specific kinds of enterprise known to us would fail. The system could not complete its tasks or achieve its purposes. Another consequence, more drastic, might be that the disturbance and disorientation would eventually ruin the internal structure of the system. The whole organized web of connections would collapse.

More important than either of these are the implications for the technological ensemble of the civilization as a whole. Ultimately, if large-scale, complex, interconnected, interdependent systems could not successfully plan, technological apraxia would become endemic. Society would certainly move to a different sort of technological development. This is clear enough to those who read Ellul, Mumford, Marcuse, Goodman, or Illich and experience horror at the critique of social existence founded on large-scale systems. What if the critics were taken seriously? What if the necessary operating conditions of such systems were tampered with? Surely society would move "backward."[4]

This idea of moving "backward" is a fascinating one. At work here is a quaint, two-dimensional, roadlike image that almost everyone (including this writer) falls into as easily as sneezing. One moves, it seems, forward (positive) or backward (negative). Never does one move upward and to the right or off into the distance at, say, a thirty-four degree angle. No; it is forward or backward in a straight line. What is understood, furthermore, is that forward means larger, more complex, based on the latest scientific knowledge and the centralized control of an increasingly greater range of variables. Hence it is clear that not to plan, not to control the circumstances of large-scale systems, is to risk a kind of ghastly cultural regression. This lends extra urgency to these measures and extra vehemence toward any criticism of the world that they produce. Surely, it is believed (and this is no exaggeration), the critics would have us *back* in the stone age.[5]

Now, everything I have said so far presupposes that large-scale technological systems are at the outset based on independent ends or purposes. It makes sense to say that technologies "serve" this or that end or need or to say that they are "used" to achieve a preconceived purpose or set of purposes. Nothing argued here seeks to deny this. In the original design, all technologies are *purposive.*

But within the portrait of advanced technics just sketched, this situation is cast in a considerably different light. Under the logic that takes one from size, interconnection, and interdependence to control and planning, it can happen that such things as ends, needs, and purposes come to be dysfunctional to a system. In some cases, the originally established end of a system may turn out to be a restraint upon the system's ability to grow or to operate properly. Strongly enforced, the original purpose may serve as a troublesome obstacle to the elaboration of the network toward a higher level of development. In other instances, the whole process that leads to the establishment of ends for the system may become an unacceptable source of uncertainty, interference, and instability. Formerly a guide to action, the end-setting process is now a threat. If the system must depend on a source that is truly independent in its ability to enforce new ends, then it faces the perils of dependency.

In instances of this kind, a system may well find it necessary to junk the whole end-means logic and take a different course. It may decide to take direct action to extend its control over the ends themselves. After all, when strongly asserted needs, purposes, or goals begin to pose a risk to the system's effective operation, why not choose transcendence? Why not treat the ends as an "input" like any other, include them in the plan, and tailor them to the system's *own needs?* Obviously *this* is the "one best way."

At this point the idea of rationality in technological thinking once again begins to wobble, for if one takes rationality to mean the accommodation of means to ends, then surely reverse-adapted systems represent the most flagrant violation of rationality. If, on the other hand, one understands rationality to be the effective, logical ordering of technological parts, then systems which seek to control their own ends are the very epitome of the rational process. It is this contrast that enables Herbert Marcuse to conclude that the technological society's "sweeping rationality, which propels efficiency and growth, is itself irrational."[6] How one feels about this depends on which model of rationality one wishes to follow. The elephant can do his dance if you ask him to, but he sometimes crushes a beautiful maiden during the performance.

Let us briefly examine some of the patterns reverse adaptation can take. Remember that as I use the term *system* here I am referring to large sociotechnical aggregates with human beings fully present, acting, and thinking. The behavior suggested, however, is meant as an attribute of the aggregate. Later I shall ask whether a change in the identities or ideologies of those "in control" is likely to make any difference.

1. *The system controls markets relevant to its operations.* One institution through which technologies are sometimes thought to be regulated is the market. If all went according to the ideal of classical economics, the market ought to provide the individual and social collectivity a powerful influence over the products and services that technological systems offer. Independent agents acting through the market should have a great deal to say about what is produced, how much, and at what price.

In point of fact, however, there are many ways in which large-scale systems circumvent the market, ways that have become the rule rather than the exception in much of industrial production. In Galbraith's version of what has become a mundane story: "If, with advancing technological and associated specialization, the market becomes increasingly unreliable, industrial planning will become increasingly impossible unless the market also gives way to planning. Much of what the firm regards as planning consists in minimizing or getting rid of market influences."[7]

Galbraith mentions three common procedures through which this is accomplished. The first is vertical integration, in which the market is superseded. "The planning unit takes over the source of supply or the outlet; a transaction that is subject to bargaining over prices and amounts is thus replaced with a transfer within the planning unit."[8]

A second means is market control, which "consists in reducing or eliminating the independence of action of those to whom the planning unit sells or from whom it buys."[9] Such control, Galbraith asserts, is a function of size. Large systems are able to determine the price they ask or pay in transactions with smaller organizations. To some extent, they are also able to control the amount sold.[10]

A third means suspends the market through contract. Here the systems agree in advance on amounts and prices to prevail in exchanges over a long period of time. The most stable and desirable of these, as we noted earlier, involve contracts with the state.

Galbraith illustrates each of these with examples from General Motors, U.S. Steel, General Electric, and others, examples I shall not repeat. The matter is now part of modern folklore. That the market is an effective means for controlling large-scale systems is known to be a nostalgic, offbeat, or fantastic utopian proposal with little to do with reality.

2. *The system controls or strongly influences the political processes that ostensibly regulate its output and operating conditions.* Other possible sources of independent control are the institutions of politics proper. According to the model, clear-minded voters, legislators, executives, judges, and administrators make choices, which they impose upon the activities of technological systems. By establishing wise goals, rules, and limits for all such systems, the public benefit is ensured.

But the technological system, the servant of politics, may itself decide to find political cures for its own problems. Why be a passive tool? Why remain strictly dependent on political institutions? Such systems may act directly to influence legislation, elections, and the content of law. They may employ their enormous size and power to tailor political environments to suit their own efficient workings.

One need only review the historical success of the railroads, oil companies, food and drug producers, and public utilities in controlling the political agencies that supposedly determine what they do and how. A major accomplishment of political science is to document exactly how this occurs. The inevitable findings are now repeated as Ralph Nader retraces the footsteps of Grant McConnell. It is apparently still a shock to discover that "regulatory" commissions are dominated by the entities they regulate.[11]

The consequences of this condition are well known. In matters of safety, price, and quality of goods and services, the rules laid down reflect the needs of the system rather than some vital, independent, and forceful expression of public interest. It is not so much that the political process is always subverted in this way; the point is that occurrences of this sort happen often enough to be considered normal. No one is surprised to find Standard Oil spending millions to fight antipollution legislation. There is little more than weariness in our discovery that the Food and Drug Administration regularly allows corporations of the food industry to introduce untested and possibly unsafe additives into mass-processed foods.[12] I am not speaking here of the influence of private organizations only. Indeed, the best examples in this genre come from public agencies able to write their own tickets, for example, the Army Corps of Engineers.

3. *The system seeks a "mission" to match its technological capabilities.* It sometimes happens that the original purpose of a megatechnical organization is accomplished or in some other way exhausted. The original, finite goals may have been reached or its products become outmoded by the passage of time. In the reasonable, traditional model of technological employment one might expect that in such cases the "tool" would be retired or altered to suit some new function determined by society at large.

But this is an unacceptable predicament. The system with its massive commitments of manpower and physical resources may not wish to steal gracefully into oblivion. Unlike the fabled Alexander, therefore, it does not weep for new worlds to conquer. It sets about creating them. Fearing imminent extinction, the system returns to the political arena in an attempt to set new goals for itself, new reasons for social support. Here a different kind of technological invention occurs. The system suggests a new project, a new mission, or a new variety of apparatus, which, according to its own way of seeing, is absolutely vital to the body politic. It places all of its influence into an effort to convince persons in the political sphere of this new need. A hypothesis suggests itself: if the system is deemed important to society as a whole, and if the new purpose is crucial to the survival of the system, then that purpose will be supported regardless of its objective value to the society.

Examples of this phenomenon are familiar in contemporary political experience and are frequently the subject of heated debate. The National Aeronautics and Space Administration faces the problem of finding new justifications for its existence as a network of "big technology." NASA has successfully flown men to the moon. Now what? Many interesting new projects have been proposed by the agency: the space shuttle, the VSTOL aircraft system, explorations to Mars, Venus, and Jupiter, asteroid space colonies. But whatever the end put forward, the fundamental argument is always the same: the aerospace "team" should not be dismantled, the great organization of men, technique, and equipment should not be permitted to fall to pieces. Give the system something to do. Anything. In the early 1960s resources were sought to fly the astronauts to unknown reaches of space; now funds are solicited to fly businessmen from Daly City to Lake Tahoe or the president to a space station for lunch.

Similar instances can be found in the recent histories of the ABM project, the SST, Boeing, General Dynamics, and Lockheed.[13] The nation may not need a particular new fighter plane, transport, bomber, or missile system. But the aerospace firms certainly need the contracts. And Los Angeles, Seattle, Houston, and other cities certainly need the aircraft companies. Therefore the nation needs the aircraft. The connections of the system to society as a whole give added punch to the effort to have reverse-adapted technological ends embraced as the most revered of national goals.

4. *The system propagates or manipulates the needs it also serves.* But even if one grants a certain degree of interference in the market and political processes, is it not true that the basic human *needs* are still autonomous? The institutionalized means to their satisfaction may have been sidetracked or corrupted, but the original needs and desires still exist as vital and independent phenomena. After all, persons in society do need food, shelter, clothing, health, and access to the amenities of modern life. Given the integrity of these needs, it is still possible to establish legitimate ends for all technical means.

The theory of technological politics finds such views totally misleading, for reverse adaptation does not stop with deliberate interference in political and economic institutions. It also includes control of the needs in society at large. Megatechnical systems do not sit idly by while the whims of public taste move toward some specifically desired product or service. Instead they have numerous means available to bring about that most fortunate of circumstances in which the social need and what the system is best able to produce coincide in a perfect one-to-one match. All the knowledge of the behavioral sciences and all of the tools of refined psychological technique are put to work on this effort. Through the right kinds of advertising, product design, and promotion, through the creation of a highly energized, carefully manipulated universe of symbols, man as consumer is mobilized to want and seek actively the goods and services that the instruments of technology are able to provide at that moment. "Roughly speaking," Ellul observes, "the problem here is to modify human needs in accordance with the requirements of planning."[14]

If the system were truly dependent upon a society with autonomous needs, if it were somehow forced to take a purely responsive attitude, then the whole arrangement of modern technology would be considerably different. In all likelihood there would be fewer such systems and with less highly developed structures. Autonomous needs are in this sense an invitation to apraxia. Adequate steps must be taken to insure that wants and needs of the right sort arrive at the correct time in predetermined quantity. In Ellul's words: "If man does not have certain needs, they must be created. The important concern is not the psychic and mental structure of the human being but the uninterrupted flow of any and all goods which invention allows the economy to produce. Whence the measureless trituration of the human soul, the true issue of which is propaganda. And propaganda, reduced to advertising, relates happiness and a meaningful life to consumption."[15]

The point here raises an important issue, which neither the apologists nor the critics of the technological society have addressed very well. Assumed in most of the writing is the continued growth of human wants and needs in response to the appearance of new technological achievements. With each new invention or innovation it becomes possible to awaken and satisfy an appetite latent in the human constitution. Potentially there is no limit to this. A want or need will arise to meet any breakthrough. But precisely how this occurs is never fully elucidated. Apparently the human being is by nature a creature of infinite appetite.

But even thinkers who believe this to be true are sometimes sobered by its implications. What if all the wrong needs are awakened? Marxists grapple with this dilemma in their analyses of "false consciousness" and "commodity fetishism" trying to explain how the proletariat should have taken such a serious interest in the debased consumer goods and status symbols of bourgeois society. Much of the neo-Marxian criticism of the Frankfurt school—the writings of Adorno, Horkheimer, Marcuse, Habermas, and others—focuses on the corruption of Marx's vision of human fulfillment in technological societies.[16] Persons in such societies certainly do lead lives of great material abundance, as predicted by Marx's theory. But the quality of their desire and of their relationships to material things is certainly not what the philosopher had in mind.

Many of Ellul's lamentations, similarly, come from his conclusion that man, or at least modern man, is indeed infinitely malleable and appetitive and, therefore, an easy mark. There is nothing that a well-managed sales campaign cannot convince him to crave with all his heart. Manipulated by "propaganda," the sum total of all psychological and mass media techniques, man wildly pursues a burgeoning glut of consumer products of highly questionable worth.[17]

It is incorrect, however, to say that needs of this sort are false. Persons who express the needs undoubtedly have them. To those persons, they are as real as any other needs ever experienced. "False" is not a response to someone who says he absolutely must have an automatic garage door opener, extra-dry deodorant for more protection, or air-flow torsion-bar suspension. No; the position of the theory is not that such needs are false but rather that they are not autonomous. A need becomes a need in substantial part because a megatechnical system external to the person needed that need to be needed.

A possible objection here is that human needs are always some variant of what is available at the time and generally desired by the society. Personal needs do not exist independent of the social environment and specific state of technics in which they occur. That is undoubtedly correct. Nevertheless, it is clear that the degree of conscious, rational, well-planned stimulation and manipulation of need is now much greater than in any previous historical period. Systems in the technological order are able to engender and give direction to highly specific needs, which in the aggregate constitute much of the "demand" for products and services. The combined impact of such manipulation produces a climate of generalized, intense needfulness bordering on mass hysteria, which keeps the populace permanently mobilized for its necessary tasks of consumption.

True, other cultures at other times have been as effective in suppressing need for religious or purely practical reasons. But this fact merely gives additional focus to the peculiar turn that a culture based on high-tech systems has taken.

5. *The system discovers or creates a crisis to justify its own further expansion.* One way in which large systems measure their own vitality

is on the scale of growth. If a system is growing, it is maintaining its full structure, replacing worn-out parts, and expanding into new areas of activity. Thinking on this subject has become highly specialized, but the basic maxim is still simple: healthy things grow.

There are times, however, when a system may find that its growth has slowed or even stopped. Even worse, it may discover that its rationale for growth has eroded. Public need for the goods or services the system provides may have leveled off; social and political support for expansion may have withered. In such cases the system has, from its own point of view, failed in its very success.

But the system is not helpless in this predicament. It does control its own internal structure, and it has command of a great deal of information about its role in society. With a little care it can manipulate either its own structure or the relevant information to create the appearance of a public "crisis" surrounding its activities. This is not to say that the system lies or deceives. It may, however, read and publicize its own condition and the condition of its environment very selectively. From the carefully selected portrait may come an image of a new and urgent social need.

Two scenarios of this sort have become familiar in recent years: the threat and the shortage. Under the psychology of the threat, the system finds an external and usually very nebulous enemy whose existence demands the utmost in technological preparation. Foreign military powers and crime in the streets have been traditional favorites. Statistics are cited to demonstrate that the enemy is well armed and busy. Society, on the other hand, is asleep at the switch, woefully bereft of tools and staff. The only logical conclusion is that the relevant system must, therefore, be given the means necessary to meet the threat as soon as possible.

The Department of Defense, to cite one noteworthy example, keeps several such plot lines in various stages of preparation at all times. If public or congressional interest in new projects or impressive hardware fails, the latest "intelligence" is readily available to show that a "gap" has appeared in precisely the area in question. This practice works best when defense systems on two continents are able to justify their growth in terms of each other's activities. Here one can see first-hand one of nature's rarities: the perfect circle.

If well orchestrated, the shortage can be equally impressive. Here the system surveys the data on its own operations and environment and announces that a crucial resource, product, or service is in dangerously short supply. Adequate steps must be taken to forestall a crisis for the whole society. The system must be encouraged to expand and to extend its sphere of control.

It may well be that there is a demonstrable shortage. What is important, however, is that the system may command a virtual monopoly of information concerning the situation and can use this monopoly for self-justification. Persons and groups outside usually do not have access to or interest in the information necessary to scrutinize the "need" in a critical way. This allows the system to define the terms of the "shortage" in its own best way. Outsiders are able to say, "Yes, I see; there is a shortage." But they are usually not prepared to ask: What is its nature? What are the full circumstances? What alternatives are available? Thus the only response is, "Do what is necessary." A number of "shortages" of this kind have been well publicized of late. There are now "crises" in natural gas, petroleum, and electricity, which a guileless public is discovering from predictable sources of information. In this instance, as in others, the almost inevitable outcome may well be "crisis = system growth supported by a huge public investment," all with a dubious relation to any clearly demonstrated need.

Some of the more interesting cases of reverse adaptation combine the above strategies in various ways. Numbers 1 and 2 as well as 2 and 3 could be anticipated as successful pairings. A particularly ironic case is that which brings together numbers 4 and 5, the propagation of need and the discovery of shortage. This is presently a popular strategy with power companies who spend millions advertising power-consuming luxury appliances while at the same time trumpeting the dire perils of the "energy crisis." Such cases might be called double reverse adaptations.

I am not saying that the patterns noted are universal in the behavior of megatechnical systems. It may also happen that the traditionally expected sequence of relating ends to means does occur, or some mix may take place. The hypothesis of the theory of technological politics is that as large-scale systems come to dominate various areas of modern social life, reverse adaptation will become an increasingly important way of determining what is done and how.

I am not arguing that there is anything inherently wrong with this. My point is that such behavior violates the models of technical practice we normally employ. To the extent to which we employ tool-use and ends-means conceptions, our experience will be out of sync with our expectations.

The Technological Imperative and the State

We are now prepared to summarize the important differences between traditional notions and the theory of technological politics.

In the traditional interpretation society has at its disposal a set of technological tools for the achievement of consciously selected social ends. Megatechnical systems are seen to be responsive and flexible. At the command of the society or its political institutions, the needed goods and services are produced. Control is one-directional and certain, leading from the source of social or political agency to the instrument.

In the technological perspective megatechnical systems are seen to have definite operational imperatives of their own, which must be met. Society stands at the disposal of the systems for the satisfaction of their requirements. The systems themselves are anything but responsive and flexible. Their conditions of size, complexity, and mutual interdependence give them a rigidity and inertia difficult to overcome. Rather than respond to commands generated by political or social processes, such systems produce demands society must fulfill or face unfortunate consequences. Confronted with these imperatives—the system's need to control supply, distribution, and the full range of circumstances affecting its operations—the immediate and expressed needs of society may seem capricious. Frequently, therefore, requirements of successful technological performance mean that control must be exercised over agencies that were formerly themselves in control.

Here, then, are two views with radically differing assertions and implied predictions. One conceivable test might be to determine how much of the U.S. Federal budget or yearly consumer dollar reflects one or the other of the processes described by the models. But although it is important to discover which of these visions best corresponds to reality, it is not my present purpose to perform such tests. Rather, I want to continue following the logic and implications of the viewpoint now unfolding before us. This will be interesting, for at this juncture there is a hitch.

All along it has been my aim to take as a basis premises and observations found in an existing body of literature and to probe, extend, and refine the arguments as I found necessary. With the idea of reverse adaptation well sketched, however, we now find ourselves at a point of major disagreement with the sources from which the present theory is drawn. The literature insists—indeed this is one of its major thrusts—that the movement of advanced technology is universally centralizing and that this centralizing tendency eventually culminates in control by an extremely powerful, technologically oriented state. But on the basis of what we have already seen, this is not a tenable conclusion.

The idea of centralization is often taken as an abbreviation of the tendencies we noticed in the previous section. It is a way of summarizing the process that leads to systematic planning and control in megatechnical organizations. It is, unfortunately, in most cases a very foggy notion. The writing on the subject (of whatever tone) employs "centralization" as if its meaning were part of everyone's innate knowledge. Seldom is the term well explicated. As best I am able to tell, critics of the technological society use it to suggest the following:

1. *That a single system or very small number of similar systems control the production and/or distribution of goods and services in a given area.* This is understood to be a consequence of technical rationality, efficiency, and the economies of scale. Under such influences one would expect to find that there would be fewer and fewer self-contained

operating units in areas of advanced technological performance over a given period of time. Formerly independent systems tend to coalesce to form large, more comprehensive networks. Typical examples can be found in the history of steel and automobile manufacturing, mass communications, railroads, modern agriculture and food distribution, and others.[18] Both Mumford[19] and Ellul cite instances from the generation and transmission of electricity. Ellul observes:

> Electrical networks may remain for some time independent of one another. But this situation cannot last when it is found that independence gives rise to general costs of no inconsiderable magnitude, difficulties in arranging the courses of the lines, and even practical difficulties in electrical technique. The interconnection of electrical networks is demanded by all technical men.[20]

2. *That within an organized system, control originates in a central source and emanates outward through the other parts.* According to the theory, the interdependence of parts in modern systems necessitates precise coordination, control, and planning of a centralized sort. Again, this is a function of the stage of development at which the particular technology stands. In less well-developed, less thoroughly systematized technological activities there is room for a high degree of independence and self-direction in similar functioning parts. But with a few exceptions, the massive ensembles of men and apparatus of the twentieth century do not permit such fragmentation. Control is exercised through centralized, hierarchical channels, which grant little real independence of decision or action at the periphery.[21]

3. *That all megatechnical systems eventually come to be controlled by a single superior center: the state.* Modern technology reaches its highest stage with the total interconnection and integration of all parts. Separate systems face problems in the control of elements outside their boundaries, often problems in their relationships with each other. The only relief from such difficulties is the mediation of the state, the supreme civil power within society. The state can guarantee the necessary conditions of operation for the separate

systems by manipulating taxation, procurement, regulations, and so forth. It can also forge and manage the links between the systems. In so doing, according to the argument, the state assumes control of all the activities in the technological order. Ellul describes the progression as follows:

> The technical "central" is the normal expression of every application. A coexistence of these centrals is implied: a completely centralized organization which ultimately encompasses all human activities. . . . Each of the centralized bodies must be put into its proper position and relation with respect to the others. This is a function of the plan, and only the state is in a position to supervise the whole complex and to co-ordinate these organisms in order to obtain a higher degree of centralization.[22]

A great deal of energy in contemporary social science has been expended in attempts to disprove claims of this sort, especially the last two. Organization theory and research have attacked the assumption that centralized control is necessary in and among complex organizations. Students of informal structure have shown that even where a centralized, hierarchical formal structure exists, a decentralized social process may carry the true weight of the enterprise.[23] Research on organization communication has shown that decentralized segments of organizations use the information they possess as a resource to further their own specific purposes and thwart control by a central source.[24] Again and again, scholars have tried to demonstrate that Max Weber's model of centralized, hierarchical bureaucracy does not describe the actual structure or behavior of many kinds of productive organization.[25] Particularly in work that involves highly sophisticated expertise and the quest for results that cannot be determined precisely in advance, hierarchy and centralization are counterproductive.[26] Critics of the "synoptic conception" of decision making have argued that centralized control is not and perhaps cannot be an effective way of managing activities within and among large-scale organizations.[27]

Evidence of this sort is sometimes taken as cause for great celebration. It suggests that the movement of efficient, effective organizational technique leads toward the decentralization of power and authority. The drift of technological history takes a favorable turn and we are given utopia, or at least a much better world, automatically. Some have trumpeted this as the "end of bureaucracy." Others have announced that based on the need for the wide dispersion of high-quality information in a complex, rapidly changing world, "democracy is inevitable."[28] If the new organization theory had been present at the time of the French Revolution, we would likely have seen a remarkable invention: the self-storming Bastille.

Obviously these findings must be taken with a grain of salt. The centralized, hierarchical organization is, in the main, still understood to be the norm by which the exceptions are measured. There may indeed be inherent limitations upon centralized decision making and control, but this does not change the fact that most of the influential organizations in any given advanced technological society still operate on that model and try, whenever possible, to see that the model succeeds. Within the majority of such large-scale organizations, public and private, there is no apparent stampede to proclaim decentralization the wave of the future. Whatever else it may be, "democracy" is not inevitable.

For the project at hand, however, my disagreements with those who predict total centralization are not based on these considerations. Instead, they rest on a different interpretation of the direction in which the theory of technological politics properly leads. The centralist conception follows the advance of technical interconnection through successively higher stages. It is correct in seeing that if this progression continued uninterrupted, it would result in the complete integration of all major parts. The hypothesis also accords with the solution I noted for overcoming problems of technological interdependence. But it fails to notice that the *stage immediately before total interconnection* poses a powerful barrier to this final step. A multiplicity of reverse-adapted, large-scale systems would, in all likelihood, have the inclination and power to oppose comprehensive centralization.

What we encounter here are different ways of thinking about

problems of the technological state and technocracy. It is worthwhile to examine them further.

The standard notion holds that technocracy is the result of two major historical trends: (1) the movement toward an increasingly specialized division of labor which eventually produces a group of supreme technicians whose specialty is organizing the works of large sociotechnical aggregations and (2) the movement toward centralized control in the technological ensemble as a whole. Whatever the ostensible form of the political order, therefore, all real control in society is exercised by an elite, often inconspicuous, acting at the central points of planning and decision. Ellul's discussion can be taken as typical. "The basic effect of state action on techniques is to co-ordinate the whole complex. The state possesses the power of unification, since it is the planning power par excellence in society. In this it plays its true role, that of co-ordinating, adjusting, and equilibrating social forces. . . . It integrates the whole complex into a plan. Planning itself is the result of well-applied techniques, and only the state is in a position to establish plans on the national level."[29] Ellul argues that the modern state is best seen as a "cold impersonal mechanism that holds all sources of energy in its hands."[30] At other times he pictures it as something of the equivalent of the carrier RNA molecule in protein synthesis; the state arranges the various parts of the technological order in their final and most successful configuration. The state is "the relational apparatus which enables the separate techniques to confront one another and to co-ordinate their movements."[31]

Necessary for the building of this supernetwork is an elite of highly trained technicians, planners, and managers, masters of "state techniques," who bring the whole into fine tune. "State constitutions do not alter the use of techniques, but techniques do act rather rapidly on state structures. They subvert democracy and tend to create a new aristocracy . . . there is a limited elite that understands the secrets of their own techniques, but not necessarily of all techniques. These men are close to the seat of modern governmental power. The state is no longer founded on the 'average citizen' but on the ability and knowledge of this elite."[32] Ellul's two chapters "Technique and the Economy" and "Technique and the State" describe in detail the various skills and training—new methods of measurement, prediction, planning, and control—which members of the new elite employ. As we would expect from our earlier encounter with technocratic theory, both citizen and politician are pretty well eclipsed by this new collection of rulers. More important, however, they are eclipsed by the circumstances in the technological order itself—the need for planning, coordination, comprehensive integration—which make the technical aristocracy necessary. "The average man," Ellul writes, "is altogether unable to penetrate technical secrets or governmental organization and consequently can exert no influence at all on the state."[33] "When the expert has effectively performed his task of pointing out the necessary ways and means, there is generally *only one logical and admissible solution* [emphasis added]. The politician will then find himself obliged to choose between the technician's solution, which is the only reasonable one, and other solutions, which he can indeed try out at his own peril, but which are not reasonable."[34]

In a summary of his position on the matter Ellul notes that there are certain connections that result "from the internal necessity of the regime." This means that all persons in positions of power must "(a) take a firm hold on the economy, (b) manage it on the basis of exact mathematical methods, (c) integrate it into a Promethean society which excludes all chance, (d) centralize it in the frameworks of nation and state (the corporate economy today has no chance of success except as a state system), (e) cause it to assume an aspect of formal democracy to the total exclusion of real democracy, and (f) exploit all possible techniques for controlling men."[35]

It is basically this outlook that has guided research into the problem of technocracy. To conduct an empirical study has meant that one find the center or centers and determine whether technical elites are actually in positions of control. These are matters that can be operationalized and tested. As we noted earlier, the studies of Meynaud and others have produced equivocal findings. Some evidence supports the hypothesis. Much of it does not.[36]

On the basis of the revised notions of technological politics presented here, however, I want to suggest an alternative conception of technocracy. It is one which avoids the pitfalls of the centralist-elitist notion and which, I believe, comes closer to defining the crucial problem that writers on this question have been aiming at for some time. The conception, although difficult to formulate in terms of precise indexes of measurement, could certainly be tested.

I offer it as follows. *Technocracy is a manifestation of two influences upon public life, which we have dealt with at some length: the technological imperative and reverse adaptation as they appear to a whole society with the force of overwhelming necessity.* From this point of view it matters little who in specific obeys the imperative or enacts the adaptation. It is of little consequence what the nature of the education, technical training, or specialized position of such persons may be; indeed, they need not be technically trained at all. Similarly, the issue has little to do with whether there is a "center" of decisions, who occupies it, or what their personal, professional, or class interests may be. The important matter concerns the kinds of decisions that any intelligent person in a position of power and authority would be required to make when confronted with accurate information on the condition of the technological order at a particular point in its development. These decisions, involving substantial commitments of society's resources, are not necessarily based upon the inherent desirability of consciously selected, independent ends. Rather, such decisions cope with necessities arising from an existing configuration of technical affairs.

Technocracy, in this interpretation, is a label properly applied to public deliberations about technology in which our traditional ends-means, tool-use notions no longer account for what takes place. The influence of socially necessary technical systems begins to constrain rather than liberate political choice. Technological imperatives appear in public deliberations as generalized "needs" or "requirements"—for example, the need for an increasing supply of electrical power—which justify the maintenance and extension of highly costly sociotechnical networks. Reverse adaptation takes the form of more specific goals and projects—the endless crusade for a new manned bomber, for instance—of the sort I have just described. One can assume that each of the technologies in question—systems of communication, energy supply, transportation, industrial production—was originally founded upon some widely accepted purpose: the accomplishment of a particular goal or the continuous supply of a product or service. But the means to the end, the system itself, requires its own means: the resources, freedom, and social power to continue its work. It needs, among other things, an atmosphere of laws and regulations to facilitate rather than limit its ability to act. The pursuit of means for the means—the provision of resources and enabling rules—may eventually lead the society to make decisions and take action far removed from its original purpose. At times these decisions can be onerous. They may promote a state of affairs that, although not in itself desired, constitutes a needed step in the development of a technical network society is committed to support.

Decisions made in the context of technological politics, therefore, do carry an aura of indelible pragmatic necessity. Any refusal to support needed growth of crucial systems can bring disaster. The alternatives range from utterly bad service, at a minimum, to a lower standard of living, social chaos, and, at the far extreme, the prospect of lapsing into a more primitive form of civilized life. For this reason, technological systems that provide essential goods and services—electricity, gas, water, waste disposal, consumer goods, defense, air, rail and automobile transportation, mass communications, and so on—are able to make tremendous demands on society as a whole. To ignore these demands, or to leave them insufficiently fulfilled, is to attack the very foundations on which modern social order rests.

Illustrations of technological politics at work can be found in the social history of virtually any significant modern technology. Through their vast, generalized power, the railroads in the nineteenth century were able to shape many public goals and institutions, including the institution of property itself. In our own time the influence and privilege of the petroleum industry offers evidence of much the same kind. How far, after all, is society prepared to alter its commitments in the quest for fuel and lubrication? The answer seems obvious: very far indeed. Footprints of technological imperatives and of reverse-adapted social ends are to be found in the extremely mundane settings—tax laws, government subsidies, right-of-way legislation, federal and state budgets and contracts, foreign policy, and the host of regulations and procedures dealing with safety, health, and the quality of goods and services in important industries and utilities.

There is, of course, an element of corruption here. As I write, the governments of Japan and the Netherlands have been thrown into chaos with the revelations that the Lockheed Aircraft Corporation had paid millions of dollars in bribes to officials of those nations to see to it that Lockheed's airplanes found an adequate market. Listening to debates in the chambers of government previous to these disclosures, however, one would never have suspected that the purchase of the planes was anything other than an expression of freely chosen public purpose. Whether in patently corrupt or more ordinary, day-to-day manifestations, the influence of technological politics tends to erode the integrity of processes through which modern society charts its course. We recall that the administration of Bacon's *New Atlantis* had an unsavoriness linked to hints of bribery. Perhaps we are discovering what the philosopher had in mind.

One virtue of the present interpretation is that technocratic phenomena are not seen as anything extraordinary. Indeed they are simply part of normal politics. To obey the imperatives or to work for reverse-adapted ends comes to be seen as the epitome of political realism. Truly tough-minded political actors anchor their ships at this dock. They know what is undeniable, which steps must be taken to maintain existing technologies, avoid catastrophe, and keep the society stable and prosperous.

This sort of political realism, furthermore, need *not* be borne by a special technocratic elite, the experts in white coats of the standard image. To the contrary, it can be handled by laymen politicians or, when the opportunity arises, even the public itself. Technocracy is rooted in the acknowledgment of conditions whose force (if not whose specific technical character) can be appreciated without years of expert training. One need not look to the Wernher von Brauns, Robert McNamaras, or Glenn Seaborgs of the world to see technocratic influences as a part of decision making. Senators, congressmen, and voters are more than capable of recognizing the essential agenda—seen in terms of jobs, the health of the economy, the continued supply of desired commodities, and so forth—and arriving at the appropriate conclusions.

Finally, the view I am taking here does not find the essence of technocracy at the center of all centers. If what we have seen is correct, then one would expect a *dispersion of power* into the functionally specific large-scale systems of the technological order. The sytems do on occasion appeal to the central decision-making organs of the state for support and assistance, but it is incorrect to say that in so doing they necessarily yield control of their affairs. Their tendency is, in fact, to resist the final centralization Ellul predicted. The direction of governance by technological imperatives and reverse adaptation runs *from* megatechnical systems *to* the state.

How do these conclusions square with our previous discussion of technocracy? An earlier chapter examined theories arguing that normal politics would be eclipsed by the rise of technical elites. We found that neither of two recent analysts of the matter, Don Price and John Kenneth Galbraith, was able to salvage a distinctive role for political knowledge and political actors in an age of advanced technical means. While their analyses do not point to the existence of a single, cohesive, self-conscious technical elite, they do point to an increasingly dominant manner of enfranchisement based on the authority of scientific and technical knowledge. The approach we have now taken traces the fundamental source of important decisions in matters involving technology beyond the role of any particular class or elite—technical, scientific, administrative, or political—to the configuration of technological conditions themselves. We would expect that both members and nonmembers of Galbraith's technostructure and all four branches of Price's scientific estate would have to respond to essentially the same imperatives.

In this understanding, the position of the traditional political actor is not eliminated. Neither is it accorded an elevated standing. Technical elites, at the same time, are not completely free to use their special knowledge in the unlimited pursuit of self or class interest. They too are bound by the constraints and inertia of the technological systems in which they work. It is true that the possession of technical knowledge does give one an advantage—an enhanced legitimacy in what one says, an opportunity for a stronger influence upon important decisions, and a chance for a greater share of material and social rewards. But this condition in itself does not entail the total supremacy of a new ruling class.

The theory of technological politics, then, is not at heart an elite theory at all. Its emphasis rests upon an even broader vision of how modern technology influences public life. The full range of technological circumstances in society tends to establish the central agenda of problems politics must confront. It also determines, to a great extent, the nature of the solutions to those problems well in advance of any real act of political deliberation.

To be commanded, technology must first be obeyed. But the opportunity to command seems forever to escape modern man. Perhaps more than anything else, *this* is the distinctly modern frustration.

The Revolution and Its Tools

The analysis offered here does not pretend to be the alpha and omega of all contemporary political knowledge. I do believe, however, that it contains some interesting insights.

As compared to more familiar models of politics—for example, those of contemporary social science or of Marxism—the theory outlined offers an alterative conception of the *motives* standing behind many dominant courses of political action. Orthodox social science would have us look to the various organized interests at work in the political arena and the means they employ to actualize their aims. Decisions and policies result from the clash of competing interests—the victories, losses, and accommodations which their actions produce. That, in sum, is politics, the social scientist explains.

Marxist analysts, on the other hand, would insist that we look at divisions of class and the manifestations of class conflict. One's social consciousness and political role are determined by one's relationship to the means of production. In capitalist societies, the enforced political mode is that which benefits the ruling class. But politics can also mean the revolutionary opposition of the proletariat to this system of rule. If the revolution succeeds, the substance of political life becomes that of building socialism.

The theory of technological politics, as I have presented it, does not set out to discredit or eliminate either of these conceptions. Instead, it seeks to illuminate certain gaps or anomalies they contain. It can be used as a supplement to either the pluralist interest model or analyses focusing on class conflict, for what the theory seeks to establish is that a significant deflection and restructuring of human motives occurs when individuals approach technologies for the solutions of their problems. These deflections give a peculiar slant to the modern political agenda that neither interest-group notions nor class theories sufficiently explain. The preceding pages have tried to state what some of these deflections are and give reasons for their occurrence.

Marxists will certainly want to say that this analysis stops short of the true point. Behind the categories I have sketched lies a stark reality, the dominance of the ruling class. What I have noted as reverse adaptation or technological imperatives are actually footprints of ruling class power. Those in positions of privilege and control are able to adjust policies, decisions, and the apparent needs of society to match their own interests. They do this so deftly as to make the whole arrangement seem necessary.

The present view, in contrast, holds that no matter who is in a position of control, no matter what their class origins or interests, they will be forced to take approximately the same steps with regard to the maintenance and growth of technological means. According to the logic we have traced, the pressures are very strong in this direction, perhaps even insurmountable. Regardless of what the original motives of the persons in power—proletarian ends or capitalist—the final results of how society operates and what it produces are about the same. Since so much of the real business of all modern societies is caught up in the support and extension of large-scale technologies, the tendency to follow imperatives and yield to the requirements of major productive systems is simply taken for granted.

It is interesting to ask whether the views of the theory of technological politics and those of orthodox Marxism can ever be fully reconciled. Herbert Marcuse and others of the Frankfurt school worked with a blend of ideas from both perspectives, with mixed success. In Marcuse's later writing we find strands of reasoning from technopolitical theory combined with premises from Hegelian, Marxist, and Freudian thought. Although I have employed some of his arguments as reference points for my discussion of the technological perspective, it must be noted that Marcuse never bases his position in this tradition alone. He delves into it on occasion when the better-developed notions of other philosophies fail to speak to a particular question. In the end he comes to the distinctively Marxian conclusion that the domination of modern technology over man is constructed in ruling-class interest. "Technology," he observes, "is always a historical-social *project:* in it is projected what a society and its ruling interest intend to do with men and things. Such a 'purpose' of domination is 'substantive' and to this extent belongs to the very form of technical reason."[37] Nevertheless, Marcuse's image of the ruling class bears the distinctive stamp of technopolitical ideas. "The capitalist bosses and owners," he writes, "are losing their identity as responsible agents; they are assuming the function of bureaucrats in a corporate machine. Within the vast hierarchy of executive and managerial boards extending far beyond the individual establishment into the scientific laboratory and research institute, the national government and national purpose, the tangible source of exploitation disappears behind the facade of objective rationality."[38]

Shortly, I will try to clarify the line of demarcation that separates Marxists' views from those of technological politics. But first we must consider another possible objection, namely, that the circumstances I have described do not matter. Even if it is true that advanced technological systems do not live up to the tool-use ideal, one can still argue that *in the aggregate* they are still tool-like. They do produce desired results. The mass productive capacity of technological society provides a set of material conditions that most persons find agreeable beyond question. No one would willingly relinquish this level of material well-being. Therefore, to suggest that highly advanced technological systems become inflexible, self-propagating, and manipulative of the human agencies they allegedly serve is merely to cite minor blemishes in an otherwise rosy picture. And in the face of such boons, what difference a few flaws?

The difference, it seems to me, can be appreciated only in comparison with aims and purposes that technological politics tends to deny. In an earlier chapter I noted some of the ways in which the influence of advanced technological practice helps to empty liberalism of much of its ideal content. An even more instructive case, however, is found in the example of modern revolutionary socialism. Here the emptying, hollowing effects seem to be even more remarkable.

For the sake of argument, let us divide the ends of twentieth-century socialism into two separate categories. In the first group we will place the normal social and economic goals of any modern industrialized nation, whatever its ideological commitment. State and society strive for a high level of employment, a high growth rate, an increasing supply of desired goods and services, and, in general, a higher standard of living. Along with this usually comes the provision for a strong military defense, a goal that sometimes pushes other aims down on the list of priorities. Whether the government and the economy succeed in fulfilling these ends is widely considered sufficient grounds for judging them on the scale of success and failure. Let us call these goals those of the technological maintenance state.

In contrast to these are the original ends of revolutionary socialism, in particular those of Marxist-Leninist theory. These offer a much broader, much more ambitious set of goals for socialist politics. These goals, supposed by Marxists to be inherent in the agenda of history, include at least the following:

1. The abolition of the class structure of society.
2. The institution of worker participation and control in the affairs of the economy and the state.
3. The construction of a truly communist society in which class domination, exploitation and the division of labor itself are totally absent.

The success of these objectives rests on the transfer of ownership of the means of production from the capitalist class to the proletariat, achieved by violent revolution. Once this change is accomplished, it is understood that the other steps will follow in proper order. The victorious proletariat will know how to proceed. In brief this is the vision of things presented us in the writings of Marx, Engels, and Lenin, a vision that has sustained radicals and revolutionaries for several generations.

For all of his criticism of capitalist exploitation, however, Marx still admired the contributions of capitalism to the development of industrial civilization. The bourgeoisie had broken down the barriers that feudal society had placed in the path of this development. Early capitalists had assembled the necessary parts of a vast material and social mechanism of progress. But with the successful completion of their role, they stood as a brake upon further advancement of the means of production. If the technical and economic capacity of modern industrial society was to continue to expand, the capitalist class would have to be done away with. After the seizure of power by the proletariat there would be, Marx argued, a remarkable burst of technological development. This would be a sure sign that the revolution was underway. In Marx's words in the *Manifesto of the Communist Party:*

The proletariat will use its political supremacy to wrest, by degrees, all capital from the bourgeoisie, to centralize all instruments of production in the hands of the state, i.e., of the proletariat organized as the ruling class, and to *increase the total of productive forces as rapidly as possible* [emphasis added].[39]

Much the same conclusion is to be found in Lenin's writings in the period shortly before and after the success of the Bolshevik Revolution. In his talk "The Tasks of the Youth Leagues" (1920), he explains, "We can build communism only on the basis of the totality of knowledge, organizations and institutions, only by using the stock of forces and means that have been left to us by the old society."[40] Lenin tells his youthful audience that it will be necessary to rebuild, expand, and modernize the entire system of production. "Communist society, as we know, cannot be built unless we restore industry and agriculture, and that, not in the old way. They must be re-established on a modern basis, in accordance with the last word in science. You know that electricity is that basis, and that only after the electrification of the entire country, of all branches of industry and agriculture, only when you have achieved that aim, will you be able to build for yourselves the communist society which the older generation will not be able to build."[41] Technical modernization and economic growth, therefore, must come first. The goals of communism will have to be postponed.

In addition to his hopes for electrification, Lenin frequently called attention to what he believed to be a major scientific development. His essay "The Immediate Tasks of the Soviet Government" (1918) contains the following statement: "The Taylor system, the last word of capitalism in this respect, like all capitalist progress, is a combination of the refined brutality of bourgeois exploitation and a number of the greatest scientific achievements in the field of analysing mechanical motions during work, the elimination of superfluous and awkward motions, the elaboration of correct methods of work, the introduction of the best system of accounting and control, etc."[42] He goes on to say that "the Soviet Republic must at all costs adopt all that is valuable in the achievements of science and technology in this field. . . . We must organize in Russia the study and teaching of the Taylor system and systematically try it out and adapt it to our own ends."[43]

The questions I wish to pose (although the history of this century raises them much more eloquently) are these: Are the goals of socialist revolution all equally compatible with the development of a large-scale technological capability? Are there some favored more by this development than others? Are there any that are in all practical fact denied? More precisely, are technical means of the sort Lenin describes truly instrumental in achieving both the goals of the technological maintenance state and those ultimate ends of the communist vision?

It is not difficult to see that in the early days of any socialist regime, the objectives of the maintenance state would receive precedence. The revolution must be defended from its enemies, both internal and external. The economy of the nation must be placed on a sound footing. For this reason, one can expect that there would be a period of rapid building, organization, and rationalization designed to keep the regime afloat at all. It is not surprising that the more ambitious ends of the revolution would be set aside temporarily while the real work at hand continues.

But the question remains: What in the long run is the relationship of these steps—steps toward rapid economic growth and the rationalization of all productive activity—to the higher goals of the communist tradition? The most poignant of all attempts to deal with this issue is in Lenin's *State and Revolution.* Written on the very eve of the October Revolution, the essay ponders the probable and desirable progress of the socialist state. In so doing it hits upon many of the crucial dilemmas which have afflicted successful revolutionaries ever since.[44]

Lenin begins with an analysis that identifies the state as an instrument of class domination. The very existence of the state, he maintains, is a sign of the irreconcilability of class antagonisms.[45] By violent revolution the bourgeois state is overthrown and the means of production seized by the proletariat. Citing Friedrich Engels, Lenin refutes the anarchist doctrine "that the state should be abolished overnight."[46] Even after the socialist revolution there will still be a need for a proletarian state to suppress the bourgeoisie. Eventually, however, the proletarian state will succeed in its work and then begin the process of "withering away."[47] Lenin goes on to defend the proposition that the transition from capitalism to "classless society" or communism will require, as Marx observed, a temporary dictatorship of the proletariat. Against the "opportunists" and "petty bourgeois parties," however, Lenin argues that the problem for the revolution is "not of perfecting the machinery of the state, but of *breaking up and annihilating it.*"[48] Again, it is the bourgeois state to which he refers. This leaves Lenin with the dual task of describing the forms a reconstructed, proletarian state will assume and of explaining how such a state will eventually "wither" and give way to communist society. Since this is a matter upon which Karl Marx was notably silent, the attempt is all the more interesting.

In many ways Lenin's discussion represents the most noble, optimistic aims of the revolutionary tradition. His description of the socialist state harkens back to the experience of the Paris Commune and to Marx's analysis of the accomplishments of that great event. *State and Revolution* envisions a socialist order characterized by a high degree of worker participation and universal equality. In contrast to much of Lenin's writing of that period, the book does not emphasize the leadership or control of the Communist party. The side of Lenin fascinated with the organization of power is largely absent. He calls for "the expropriation of the capitalists, the conversion of *all* citizens into workers and employees of *one* huge 'syndicate'—the whole state—and the complete subordination of the whole of the work of this syndicate to the really democratic state of the Soviet of Workers' and Soldiers' Deputies."[49] He speaks of a time when "all members of society . . . have learned how to govern the state *themselves*, have taken this business into their own hands."[50] "It is *only* with Socialism that there will commence a rapid, genuine, real mass advance, in which first the majority and then the whole of the population will take part—an advance in all domains of social and individual life."[51]

Beyond socialism lies the eventual transition to communist society in which mankind will move "from formal equality to real equality, to realizing the rule, 'From each according to his ability; to each according to his needs.'"[52] Lenin admits that he does not know exactly how this will be accomplished. "By what stages, by means of what practical measures humanity will proceed to this higher aim—this we do not and cannot know."[53]

It is clear, nevertheless, that Lenin believes that developments in the socialist stage will be compatible with the eventual movement toward communism. He is particularly confident that the proposed rapid development of technology will facilitate and could not possibly hinder the revolution's ultimate purpose. Apparatus, technique, and organization are in his eyes nothing but tools waiting to serve humanity.[54] But if what we have seen here is correct, Lenin's hopeful anticipations can be taken as forecasts of bad weather.

A case in point can be found in Lenin's treatment of relations of authority. The book takes pains to deny the legitimacy of all capitalist and bourgeois democratic institutions. For example, it argues that existing parliaments are not representative bodies but merely means "to repress and oppress the people."[55] But while it denies the authority of all prerevolutionary political institutions, the essay is careful to defend a residual authority carried over from the capitalist world—the legitimacy of technique and of the leadership of technically proficient persons. At the same time that Lenin calls for universal equality and political participation without subordinate roles, he also argues that technology requires subordination, hierarchy, and obedience. In the following passage he cites with approval Engels's views on the matter:

> He first of all ridicules the muddled ideas of Proudhonists, who called themselves "anti-authoritarians," *i.e.*, they denied every kind of authority, every kind of subordination, every kind of power. Take a factory, a railway, a vessel on the high seas, said Engels—is it not clear that not one of these complex technical units, based on the use of machines and the ordered co-operation of many people, could function without a certain amount of subordination and, consequently, without some authority or power?[56]

From this Lenin concludes, in agreement with Engels, "that authority and autonomy are relative terms, that the sphere of their application varies with the various phases of social development, that it is absurd to take them as absolute concepts."[57]

Throughout *State and Revolution* Lenin calls attention to the authority and discipline associated with technological processes. The technique of modern industry, he observes, "requires the very strictest discipline, the greatest accuracy in the carrying out by every one of the work alloted to him, under the peril of stoppage of the whole business or damage to mechanism or product."[58] But while this discipline is necessary, it is also, he believes, entirely benign. In particular, it will *not* include the maintenance of bureaucratic administration. "The workers, having conquered political power, will break up the old bureaucratic apparatus . . . and they will replace it with a new one consisting of these workers and employees, *against* whose transformation into bureaucrats measures will at once be undertaken."[59] The measures include election, recall, equality of payment for all jobs, and a system in which each person has a turn at administrative control and superintendence.[60]

If Lenin is right, the revolution should bring a quantum jump in the technological capacity of society as well as a remarkable shift of power and control to the masses of people. Both of these developments, furthermore, will occur in approximately the same arena—the industrial organizations now owned by the proletariat. Is it possible that the two might overlap or even conflict? Lenin foresees no such difficulty. After the revolution abolishes private property and bureaucracy, all such problems are solved. It is possible, he affirms, "within twenty-four hours after the overthrow of the capitalists and bureaucrats, to replace them, in the control of production and distribution, in the business of *control* of labour and products, by the armed workers, by the whole people in arms."[61]

This sounds reassuring until Lenin adds a poignant parenthetical comment: "The question of control and accounting must not be confused with the question of the scientifically educated staff of engineers, agronomists and so on. These gentlemen work today, obeying the capitalists; they will work even better tomorrow obeying the armed workers."[62] Thus, the authority of scientific technique finds personification. Lenin asserts that men of this sort will remain entirely tool-like in their role. But is this an assurance we can accept?

Two clouds form on the horizon. First is the possibility that a new class will emerge from the group that Lenin sets aside. Will the knowledge and obvious productivity of the technically proficient elevate them to a position far above that of the armed workers? Is it not conceivable that such people will begin to demand a greater proportional reward? Is it not likely that they will organize their power as a cohesive class unit pursuing its own interest?

A second and, I think, even more important prospect lies in the possibility that the search for efficiency and productivity may find no harmony with the mode of worker participation and control. Lenin already has in mind a particular path to the optimal, scientific organization of the means of production. The path thrives on highly rationalized, authoritarian technics. What will happen if the demands and decisions of the workers point in the direction of a less efficient, rational, and productive industrial system? Lenin outlines a set of noble political ends and a set of optimal technological ends. He seems to assume that the two will have a good fit. Clearly, nothing could be less certain. In the modern world the noble and the optimal are often found in conflict.

It is not my purpose to give a full analysis of Soviet historical experience. But I believe the evidence of more than a half-century leaves us with some fairly well-established conclusions. Judged according to the aims of the technological maintenance state—productivity, efficiency, economic growth, supply of consumer goods and services, military security, and so forth—the record of the regime has gotten increasingly better. It is, in fact, a remarkable success. But if we measure its progress by the distinctive ends of the revolutionary tradition—abolition of class structure, building of worker participation and control, movement toward a truly communist social order—the Soviet Union has failed miserably. These objectives have now been for the most part simply forgotten or drained of their original meaning.

Why this happened has been the subject of decades of heated controversy, from Rosa Luxemburg to Daniel Cohn-Bendit.[63] Certainly, there are many factors which account for the course that the revolution took —the personalities of its leaders, factional conflict, the pressures of war and economic crisis, and so forth. The element singled out here—the stringent demands of the technological systems which the revolution inherited from the capitalist order or implemented soon thereafter—is, I believe, arguably one of the more important of such factors, although it would be absurd to conclude that this influence alone accounts for the tone and direction that Soviet policy took after 1917. But it *is* one aspect of the story. Its special significance becomes apparent when one considers the question, Why did the change in ownership of the means of production make so little difference in changing the basic pattern of social and economic life? The fact that one achieves *ownership* does not mean that one is immediately capable of changing the networks of apparatus, technique, and organization to suit revolutionary goals. Particularly if survival is at stake, tampering with the mechanism takes on a very low priority.[64]

Here an interesting problem for Marxist theory comes into view. Lenin, like many other revolutionary socialists, *did* believe that the contradictions we now see between the goals of the technological maintenance state and those of communist revolution would eventually be resolved. All of the steps he recommended for the rapid technological development of Russia—Taylorism, electrification, and the restoration of agriculture and industry "in accordance with *the last word in science.*"[65]—were in his eyes completely compatible with and, indeed, necessary for the advent of true communism. His conjectures about the way this would occur are based on one of the few fairly solid statements that Karl Marx ever made on the conditions that would exist in the further stages of revolutionary development. Lenin quotes Marx's *Critique of the Gotha Program:*

> In a higher phase of Communist society, when the enslaving subordination of individuals in the division of labour has disappeared, and with it also the antagonism between mental and physical labour; when labour has become not only a means of living, but itself the first necessity of

life; when, along with the all-round development of individuals, the productive forces too have grown, and all the springs of social wealth are flowing freely—it is only at that stage that it will be possible to pass completely beyond the narrow horizon of bourgeois rights, and for society to inscribe on its banners: from each according to his ability; to each according to his needs![66]

Lenin expands upon Marx's passage with an astounding claim. A "gigantic development of the productive forces," he says, "will make it possible to break away from the division of labor."[67] In particular, he understands this to mean that technology will advance to a point at which the "antagonism between mental and physical labour" can be done away with altogether. Capitalism, he explains, "*retards* this development."[68] It is the final stumbling block standing between humanity and a totally equalitarian social order. Hence the rationale for destroying capitalism is much more than a desire to rid the world of exploitation. Once the capitalists are gone, advanced technology will provide the objective conditions necessary for "that high stage of development of Communism."[69]

What was Lenin, or for that matter Marx, thinking of here? It is no small question. On it hangs the matter of what it is possible for a successful Marxian revolution to accomplish. Lenin's expectation is that all advanced, scientific technologies point toward an eventual abolition of the division of labor. On the basis of twentieth-century experience with technology, is this a reasonable expectation? Quite the contrary, it would seem. As larger, more powerful, more complex technical systems are built, the division of labor becomes more and more highly developed. There is a proliferation in the kinds of levels and compartments in that division. The gap between physical and mental tasks grows larger rather than smaller. Purely technical progress as it affects the social relations of work seems to require an increasingly refined and stratified collection of productive roles. Almost nothing in our experience shows us anything different.

It is precisely at this point that Lenin's and most other revolutionary and utopian schemes go on the rocks. One can seek the high levels of productivity that modern technological systems bring. One can also seek the founding of a communal life in which the division of labor, social hierarchy, and political domination are eradicated. But can one in any realistic terms have both? I am convinced that the answer to this is a firm "no." The reasons can be found in the theory we have examined as to what constitutes a technology, reasons confirmed by the historical evidence of this century.

Lenin, however, does not leave us totally in the dark about what his own expectations were. If we study his argument about the relationship of technology and the division of labor, we find his belief that the divisions were *already* in fact breaking down. The capitalists were, he believed, doing all they could do to enforce purely contrived separations in varieties of work, mental and physical, which would surely collapse with the coming of a revolution. The progress of capitalist technique had made the distinction between workers and administrators totally obsolete. "In its turn, capitalism, as it develops, itself creates *prerequisites* for 'every one' *to be able* to really take part in the administration of the state. Among such prerequisites are: universal literacy, already realised in most of the advanced capitalist countries, then the 'training and discipline' of millions of workers by the huge, complex, and socialised apparatus of the post-office, the railways, the big factories, large-scale commerce, banking, etc., etc."[70] Most importantly, capitalist technology had brought with it a profound simplification in all socially productive functions. It was no longer possible to defend bureaucratic and administrative roles on the basis of the extremely high levels of education and expertise required for effective performance. "The accounting and control necessary for this have been *simplified* by capitalism to the utmost, till they have become the extraordinarily simple operations of watching, recording and issuing receipts, within reach of anybody who can read and write and knows the first four rules of arithmetic."[71] In Lenin's view, therefore, the combined effect of education and simplification in the sphere of productive relations would continue to make the division of labor—as well as the hierarchical, stratified social order based upon it—a feeble anachronism doomed to extinction.

Lenin's point here does make some sense. We all know of positions of supposed expertise and technical sophistication in modern organizations that actually require little special education, extraordinary knowledge, or talent. It is also true that certain kinds of technology do simplify the tasks of manual work and of information handling. But to find in these facts a tendency toward universal simplification and the abolition of the very rationale for the division of labor is to carry the point to absurdity. The dominant tendency moves in the opposite direction. Especially if one decides to take one's cues from the most sophisticated technology available, the number of specialized functional divisions required in almost every productive enterprise will continue to increase. The linking together of these functions creates a situation of greater complexity rather than less.

It is interesting to speculate about a kind of technology that would allow the division of labor to become a thing of the past—a technology that would not breed social stratification, hierarchy, and inequalities of privilege. But this is not at all the technology Lenin describes. It is, furthermore, not very helpful to begin the search for such things by assuming that the normal course of technological progress will create them automatically.

Here, then, to add to our growing catalog, is yet another meaning of the notion of autonomous technology. The Marxist faith in the beneficence of unlimited technological development is betrayed by the fact that the destinations reached are simply the wrong ones. The supreme confidence that things would be different if only the technological tool could be owned and controlled by the workers runs into precisely the difficulties we have noticed. It is difficult to get a handle on these "tools," difficult to do anything different with them, difficult to overcome the patterns they so strongly impress upon social relationships. Thus, even when the revolution is in season, its orientation toward things technical lays it wide open to the reverse adaptation of its best ends. To the horror of its partisans, it is forced slavishly to obey imperatives left by a system supposedly killed and buried. Technological politics does away with much of the villainy in history, but it leaves the tragedy intact.

Some hopes are raised by the possibility that the Chinese revolution has learned a sobering lesson from the experience of the Russians and is trying a different route to socialism. Through the institution of the dialectical "three-in-one" concept and other devices in the organization of work, China evidently seeks to counteract technological tendencies at odds with Mao's vision of politics. Evidence of the success of this attempt, however, remains entirely inconclusive.[72]

It is important to notice that the movement I have been tracing here is not an ideological one. The victory of technics over politics does not occur because large numbers of persons have suddenly abandoned liberalism, Marxism, or some other political philosophy to adopt a new set of doctrines. It is not necessary to conclude, for example, that a "cult of efficiency," an enthusiastic group of technophiles, has taken over the reins of government around the world. Habermas argues that the practice of scientific technology does bring with it specific ideological commitments.[73] That is no doubt correct. But it would be a mistake to suppose that this ideology has been taken up by great numbers of men and women as a political cause. Indeed, it can be said that those who best serve the progress of technological politics are those who espouse more traditional political ideologies but are no longer able to make them work.

Ellul puts it well: "Things happen today in the political sphere without the benefit of the minutest theory."[74] He suggests that if one wants a label for the kind of government the mid-twentieth century has produced, the most accurate one would be "state capitalism." But, again, this has nothing to do with the rise of a doctrine or ideology of state capitalism. Neither is it a manifestation of the ascendance of the ideology of efficiency. It is the *fact* of efficiency, he argues, that really matters. "Political doctrine, since about 1914, works in this way: the state is forced by the operation of its own proper techniques to form its doctrine of government on the basis of technical necessities. These necessities compel action in the same way that techniques permit it. Political theory comes along to explain action in its ideological aspect and in its practical aspect (frequently without indicating its purely technical motives). Finally, political doctrine intervenes to justify action and to show that it corresponds to ideals and to moral principles. The man of the present feels a great need for justification. He needs the conviction that his government is not only efficient but just. Unfortunately, efficiency is a fact and justice a slogan."[75]

The position we have seen can be summarized as follows: technology is now a kind of conduit such that no matter which aims or purposes one decides to put in, a particular kind of product inevitably comes out. This state of affairs is not well suited to political theory in any traditional sense. The theory of technological politics itself, even when it hones its critical edge, usually ends up being little more than an elaborate description. Somehow one has to remember the content of other theories and visions in order to catch a sense of its significance at all. In the end, the best "theory" for a world of this kind might well be a series of aerial photographs showing the gradual expansion of the technological grid.

Chapter 7

Complexity and the Loss of Agency

We have already covered ground sufficient to raise serious questions for conventional interpretations of technology. The blithe claim that the apparatus, techniques, and organized systems of the modern age are merely neutral, available tools subject to spontaneous use and to obvious modes of control has been criticized and, I believe, badly dented. An additional subject, however, remains to be examined before the viewpoint we have been developing is complete. Missing from our treatment so far is an encounter with one of the most widely cherished beliefs in the conventional understanding—that there is nothing that human beings know so well as their own artifice.

The most eloquent of modern spokesmen for this point of view was Thomas Hobbes. A century before similar arguments were advanced by Giambattista Vico, Hobbes pointed out that the highest certainty was found in purely artificial things, things people put together and can also take apart. "Geometry, therefore is demonstrable," he wrote, "for the lines and figures from which we reason are drawn and described by ourselves; and civil philosophy is demonstrable because we make the commonwealth ourselves."[1] This view did not, of course, originate with Hobbes. In *The Republic* Plato's working image of certainty rests squarely upon examples drawn from *technē*, the practical and scientific arts. To know something thoroughly, one would know it in the sense that a craftsman, physician, or other highly skilled person knows his work and its products. Finding in this model a reliable link between theory and practice, Plato suggests that the knowledge of justice applied to the construction of the best state must have the same quality. Politics becomes an activity of enlightened artifice.[2]

In his introduction to the *Leviathan* Hobbes carries this confidence to its most optimistic conclusion. He portrays the state as a marvelous device, a well-constructed machine which mirrors God's own best little mechanism, man.

For what is the *heart*, but a *spring;* and the *nerves*, but so many *strings;* and the *joints*, but so many *wheels*, giving motion the whole body, such as was intended by the artificer? *Art* goes yet further, imitating that rational and most excellent work of nature, *man*. For by art is created that great LEVIATHAN called a COMMONWEALTH, or STATE, in Latin CIVITAS, which is but an artificial man; though of greater stature and strength than the natural, for whose protection and defence it was intended; and in which the *sovereignty* is an artificial *soul*, as giving life and motion to the whole body.[3]

Hobbes argues that, in principle at least, such a state could be constructed with absolute perfection and certainty. The artificer, in this case a single creator, the political theorist, would know in advance the complete contents of well-established order. He would know it right down to its very *soul*, the source of its life and motion.

An interesting variation of the same idea occurs in the literature of autonomous technology. Of course, we are by now accustomed to seeing society depicted as a vast mechanism. But writers concerned with problems of technology-out-of-control have frequently echoed Hobbes in suggesting that such an artifact—the Leviathan of interconnected technical systems—has a soul of its own. In its very nature, this is not a soul that is planned in advance or inserted into the machine by design. Instead, it is a quality of life and activity springing from the whole after the myriad of parts have been fashioned and linked together. A ghost appears in the network. Unanticipated aspects of technological structure endow the creation with an unanticipated *telos*.

Now, what can possibly be meant by that?

If I am not mistaken, the reader has probably read the previous chapters with mixed feelings. Some of the points I have drawn from the technological perspective may seem valid, others subject to criticism. That is inevitable. In reopening these questions, in trying to jolt a whole collection of habits in our thinking, I have deliberately courted some extremes. This intention is placed to its ultimate test as it confronts ideas of the following sort:

Technique has become a reality in itself, self-sufficient, with its special laws and its own determinations.[4]

Technique tolerates no judgment from without and accepts no limitation.[5]

> Technique, in sitting in judgment on itself, is clearly freed from this principal obstacle to human action.[6]
>
> The power and autonomy of technique are so well secured that it, in its turn, has become the judge of what is moral, the creator of a new morality. Thus, it plays the role of creator of a new civilization as well.[7]

These assertions are found in the section of Ellul's book entitled "Autonomy of Technique," a section which tries to describe the ultimate significance of technique in modern history. By now, we are prepared to understand his metaphysical formulations as broad, challenging metaphors. We can penetrate their aura of mystification and interpret them in terms of the specific issues we have examined. At the same time, one must not overlook the fact that Ellul is talking as if technique had a soul. This is more than a metaphor. There is a sense in which he means it to be taken in a literal sense. Ellul agrees with Hobbes that the vast artificial creation does have a willful, active, self-determining quality of its own. It is not the sovereignty of the commonwealth, fashioned by a single omniscient creator, but the autonomy of technique generated within the interlocking parts of a complex structure built in bits and pieces over the years by millions of intelligent hands.

To put it differently, Ellul moves to the conclusion that technique does manifest an intrinsic Geist. There is no doubt that this thesis has religious meaning for him. Man now worships Mammon rather than God. Mammon obliges by taking on lifelike characteristics. Ellul, however, goes even further, offering the idea as a factual assertion as well. We saw earlier that his sociology rests on the validity of Durkheimian "social facts" which appear, sui generis, in the social order. Extending this view he proposes as a question of fact that the modern technological ensemble develops a character, possibly even a spirit, unto itself, which is distinct from the structure or behavior of any of its specific parts.

This is, indeed, a most peculiar aspect of the style of thinking we have been studying. But one need not embrace either Ellul's religious sentiments or adopt the scientific problems in the notion of social facts to make sense of the fundamental point in question. There is another path available.

If, by comparison, one remains a skeptic and finds it impossible to believe in the existence of a human soul, one sometimes tries to explain that the notion of "soul" is simply a way of representing certain phenomena in the life of human beings not as yet well understood. "Soul" fits conveniently into a lacuna in our understanding. As the sciences teach us more about human physiology, chemistry, psychology, and behavior, we have fewer and fewer occasions in which to use the term. Similarly, the idea of autonomous technology in the grand sense that Ellul ultimately suggests can be taken as an attempt to fill another kind of gap in our understanding. Which? I am convinced it is this: *the gap between complex phenomena that are part of our everyday experience and the ability to make such phenomena intelligible and coherent.* It is this problem, in a true sense the capstone for the other issues raised so far, to which we must now turn. How well, after all, do members of the technological order know their artificial environment?

Complexity: Manifest and Concealed

It is commonly assumed that ours is an age of increasing intelligence. Man knows more about the world than at any previous time. According to some estimates, the sum of scientific knowledge is doubling with each decade. Worldwide literacy increases at a rapid rate. Schools, universities, and research institutions proliferate, making access to knowledge available to more and more people. In general, it is safe to say that most social institutions employ much more refined varieties of technical information than ever before.

Nevertheless, there is a case to be made that this is also an era of rapidly increasing ignorance. It is true that more and more knowledge is gathered through an ever-expanding array of means. Yet mastery of knowledge appears to be waning in the sense that ever less of what is known can be digested, taught, learned, or utilized by any given individual, group, or organization. If ignorance is measured by the amount of available knowledge that an individual or collective "knower" does not comprehend, one must admit that ignorance, that is, *relative ignorance,* is growing.

In *The Act of Creation,* Arthur Koestler speaks of this situation in the life of the well-educated mid-twentieth-century man.

> He utilizes the products of science and technology in a purely possessive, exploitive manner without comprehension or feeling. His relationship to the objects of his daily use, the tap which supplies his bath, the pipes which keep him warm, the switch which turns on the light—in a word, to the environment in which he lives, is impersonal and possessive. . . . Modern man lives isolated in his artificial environment, not because the artificial is evil as such, but because of his lack of comprehension of the forces which make it work—of the principles which relate his gadgets to the forces of nature, to the universal order. It is not central heating which makes his existence "unnatural," but his refusal to take an interest in the principles behind it. By being entirely dependent on science, yet closing his mind to it, he leads the life of an urban barbarian.[8]

Koestler's description of the "urban barbarian" seems to me accurate. But his explanation of the malady is inadequate. It is not that individuals refuse to learn the principles upon which their environment runs. Rather, it has become impossible for them to learn anything but the smallest portion of the knowledge necessary to make their world fully comprehensible.

The dimensions of this state of affairs can be appreciated by reexamining the question of complexity.[9] As we have seen, technical creation advances through division and resynthesis. Segments of the preexisting world are broken down and reassembled in new and more productive forms. An important characteristic of this process is that while the phenomenon of the finished whole is available to the public, the knowledge upon which it is fashioned usually does not become part of widely shared social experience. On the contrary, the detailed information and theory appropriate to a particular technique, apparatus, or organization becomes the exclusive domain of specialists in that field.

Nonspecialists cannot be expected to know anything about the subject. It would require too much time and effort for the individual to learn the workings of all the arrangements in his milieu.

One is left, therefore, with a society in which knowledge is highly fragmented, in which each person knows one or a few things very well. Its members are unable to account for the operations of parts not included in their particular specialties. They are not prepared to understand the synthesis of larger wholes. Society is composed of persons who cannot design, build, repair, or even operate most of the devices upon which their lives depend. In this sense, specialists of various stripes are left to trade on each other's ignorance.

The condition resulting from the circumstances described is one I wish to call *manifest social complexity*. The technological society contains many parts and specialized activities with a myriad of interconnections. The totality of such interconnections—the relationships of the parts to each other and the parts to the whole—is something which is no longer comprehensible to anyone. In the complexity of this world, people are confronted with extraordinary events and functions that are literally unintelligible to them. They are unable to give an adequate explanation of man-made phenomena in their immediate experience. They are unable to form a coherent, rational picture of the whole. Under these circumstances, all persons do and, indeed, must accept a great number of things on faith. They are aware that the major components of complex systems usually work, that other specialists know what they are doing, and that somehow the whole fits together in relatively good adjustment. Their way of understanding, however, is basically religious rather than scientific; only a small portion of one's everyday experience in the technological society can be made scientific. For the rest, everyone is forced to depend upon and have faith in matters about which one has little information or intelligent grasp. It is this condition that Ellul describes as the source of the modern versions of mystery, magic, and the sacred.[10]

The unintelligibility of complex sociotechnical phenomena is compounded by another set of circumstances. The development of advanced electronics has made it possible to conceal the complexity of important functions. Many kinds of interactions can be reduced to electrical impulses and placed in the microminiaturized circuits of computers and instruments of communications. Such devices are highly complex in their internal structures and processes. But this complexity is enclosed within the boundaries of the apparatus and removed from public view. Matters of intricate business formerly handled through face-to-face encounters or bureaucratic paperwork are transferred to hardware capable of astounding speed and efficiency. Banking, travel reservations, factory management, product distributions, and the day-to-day business of countless organizations are now taken care of in this manner. In another decade or so, money itself might disappear, to be replaced by instantaneous computerized accounting. Manifest social complexity is replaced by *concealed electronic complexity*. Relationships and connections once part of mundane experience (in the sense that some person had to attend to them at each step) are now transferred to the instrument. The unintelligible mass of sociotechnical interconnections is enshrouded in abstraction.

Concealed complexity gains much of its significance by contributing new dimensions to the technological extension of human capacities. Electronic media, as McLuhan tells us again and again, overcome limitations of space, time, and distance. They enable one to communicate and to act effectively over great expanses in an instant. The most important of the technical systems of the modern age are structurally complex and spatially extended and make the handling of daily business through face-to-face relations increasingly irrelevant.

In encounters with both manifest and concealed complexity, the plight of members of the technological society can be compared to that of a newborn child. At first the child cannot organize the buzzing chaos of worldly phenomena. Much of the data that enters its sense does not form coherent wholes. There are many things the child cannot understand or, after it has learned to speak, cannot successfully explain to anyone. But with time children begin to make sense of the world. Through trial and error and processes of social learning, they begin to comprehend and gradually master more and more of the environment in which they live.

Citizens of the modern age in this respect are less fortunate than children. They never escape a fundamental bewilderment in the face of the complex world that their senses report. They are permanently swamped by extraordinary phenomena bombarding them from every side. They are not able to organize all or even very much of this into sensible wholes. Unlike the child, their personal probings and social learning offer no breakthrough into comprehensive understanding. The best anyone can do is to master a few things in the immediate environment—enough for job, family, and a roof over one's head. Anything beyond that is both very difficult and risky. Unlike the child who when successful in his learning reaches farther, members of technological society learn to play it safe, to erect limits and abide by them. Indeed, this is something of a prophylactic tendency, for to encompass the whole (or even a substantial slice) would be to risk madness. Human sensibility, understanding, and courage simply do not reach far enough. Information overload, the human equivalent of the blown fuse, awaits those who attempt to test this fate.

An objection might be raised that such conditions of ignorance do not constitute a problem. The specialized skills and complex systems of the technological society are available to anyone with access to them regardless of how much the person knows about such things. All one needs to know is the simple matter of who or what to employ and where to plug in. There is no *need* to understand electricity or plumbing in order to turn on a garbage disposal. No comprehensive grasp of airlines organization is necessary to fly United. If anything, according to this view, the lack of detailed information and the lack of a capacity to make crucial complexities intelligible are positive virtues. After all, why bother? Why not use the end products and let it go at that? Why take on the extra burden of having to understand each instance of sophisticated artifice one comes upon? Obliviousness to such things is, in fact, liberating. It permits us the time to lead lives which encompass a variety of activities in work, travel, communication, and leisure of a scope totally unmatched in previous history. Potentially, at least, a world of this sort opens these opportunities to all persons. *Access* is really all that is required.

Another possible objection might be that there is no problem because difficulties of the sort I have mentioned either do or soon will have remedies. Society will find means of effective synthesis to deal with the problems of understanding that arise from an increasingly complex milieu. Systems theory, artificial intelligence, or some new distinctively modern way of knowing will alleviate the burdens of information overload.

Both of these responses have shortcomings. The first view is essentially a reaffirmation of the image of hyperactive life in the technological society. It excuses ignorance by pointing to the possibility of a new level of performance. Members of the society are able to do more things, more efficiently, over farther distances, at much faster speeds. The busyness of the daily everything enhances freedom. It is, furthermore, what most people want. They are persons "on the go" with "all systems go." *More, farther,* and *faster* is the formula for virtue in the modern age, our frenetic equivalent of the *areté* of the Greeks or the piety of the Puritans.

But the hyperactive image belies a remarkably passive substance. Yes, the available technologies do enable members of the society to do a great many things. But the operators of modern gadgetry are almost totally docile with respect to exercising any determining influence upon the design, implementation, day-to-day operation, or choice of outputs of the systems that surround them. Even their notion of active, vital use is more accurately seen as a passive utilization—a totally accepting, unquestioning relationship with technologies over which they have no real power. This is, after all, what is involved in the variety of citizenship that defines participation as consumption. One accepts uncritically what issues from the productive mechanism.

The current furor of the consumer movement confirms this state of affairs. Militant consumers have at last begun to ask for their money's worth. They insist on quality, health, and safety from the goods they purchase. But deep in the foundations of their argument is the conviction that if producers truly were responsible, society would not need watchdog measures. Producers would offer adequate goods as a matter of course. We can perceive a yearning to return, after the battle of the supermarket, to the tranquil time when the products were so fine that one need not care how they were made. The idea that one might want to pay much closer attention to the technological processes influencing one's existence and that this might be an area for critically informed action is still totally foreign.

The desire for access to the "black boxes" produced by technology, therefore, does not imply a desire for access to the inner workings of technology itself. One becomes accustomed to the idea that systems are too large, too complex, and too distant to permit all but experts an inside view.

In answer to the second objection—that there now or soon will exist tools of intellectual synthesis capable of dealing with extreme complexity—I must report that I have found no such tools in practice. I have elsewhere surveyed the various candidates for this honor—systems theory and systems analysis, computer sciences and artificial intelligence, new methods of coding great masses of information, the strategy of disjointed incrementalism, and so forth.[11] As relief for the difficulties raised here—the bafflement of human intelligence by the sociotechnical complexity of the modern age—none of these offers much help. This is not to say that these tools are ineffective in performing certain tasks. They are effective in certain respects. But the hope that such tools and methods would be able to unify fragments of human knowledge and give a more intelligible picture of the whole appears frustrated. We still suffer from a situation that Amitai Etzioni has labeled "high input" with "deficient synthesis."[12]

One illustration will make clear what is involved in the search for intellectual tools of this sort and also what is involved in the failure to find them. For many years Ludwig von Bertalanffy and his colleagues have actively sought a "general systems theory." The problem to be overcome is similar to that which I have just described. "Modern science," he explains, "is characterized by its ever-increasing specialization necessitated by the enormous amount of data, the complexity of techniques and of theoretical structures within every field. Thus science is split into innumerable disciplines continually generating new subdisciplines. In consequence, the physicist, the biologist and the social scientist are, so to speak, encapsulated in their private universes, and it is difficult to get from one cocoon to the other."[13] Bertalanffy's plan to reunify scientific knowledge is founded on both a metaphysical supposition and faith in a particular kind of technique. He begins with the idea that the world is stratified into distinct system levels, which differ from each other in detail but share "general system properties." Scientists in various fields would study specific physical systems as things in themselves and develop mathematical models to describe their behavior. The models could then be compared, thus revealing isomorphisms of "structural uniformities of the schemes we are applying." "We can ask,"he suggests, "for principles applying to systems in general irrespective of whether they are physical, biological or sociological in nature. If we pose this question and conveniently define the concept of system, we find that models, principles and laws exist which apply to generalized systems irrespective of their particular kind, elements and forces involved."[14]

It is perhaps premature to judge the final success or failure of this ambitious program. But in the introduction to a recent book Bertalanffy speaks of his disillusionment about the progress of the movement. He notes that today "the student in systems science receives a technical training which makes systems theory—originally intended to overcome current over-specialization—into another of hundreds of academic specialties. Moreover, systems science, centered in computer technology, cybernetics, automation and systems engineering, appears to make the systems idea another—and indeed the ultimate—technique to shape man and society ever more into the 'mega-machine' which Mumford has so impressively described in its advance through history."[15]

Bertalanffy's dream of an ultimate intellectual tool—a science of sciences—is frustrated by the very conditions it hoped to conquer. The cure becomes merely another manifestation of the disease.

This appears generally true. The only effective way of dealing with the complexity of the world and the fragmentation of knowledge is to carry the division and specialization ever further. There are still those who yearn for a new synthesis, a comprehensive vision of natural and man-made reality. But the mainstream knows such persons to be futile generalists, mystics, or crackpots, not to be taken seriously. In terms of the real business of the culture, Paul Valéry's conclusion in the essay "Unpredictability" (1944) still seems accurate:

> What has happened? Simply that *our means of investigation and action have far outstripped our means of representation and understanding.*
>
> This is the enormous *new fact* that results from all other *new facts.* This one is positively *transcendent.*[16]

It was precisely this problem that brought H. G. Wells, the most optimistic of the prophets of modernity, to final despair. As we have seen, Wells struggled for decades to come to grips with the implications of a world of advanced technological marvels and to compose an idea of a better world from them. But he finally concluded that all such attempts were vain. In his last published work, *The Mind at the End of Its Tether* (1945), he argues that the human mind is no longer capable of dealing with the environment it has created.

> Spread out and examine the pattern of events, and you will find yourself face to face with a new scheme of being, hitherto unimaginable by the human mind. This new cold glare mocks and dazzles the human intelligence, and yet, such is the obstinate vitality of the philosophic urge in the minds of that insatiable quality, that they can still under its cold urgency seek some way out or round or through the impasse.
>
> The writer is convinced that there is no way out or round or through the impasse. It is the end.[17]

Fearing the consequences of the wrong kind of integration of the mass of new parts, Wells observes: "It is also possible that hard imaginative thinking has not increased so as to keep pace with the expansion and complication of human societies and organizations. That is the darkest shadow upon the hopes of mankind."[18]

But is a comprehensive vision of the complicated whole really necessary? A major argument of contemporary social science seeks to demonstrate that society is able to operate quite well without this capacity. Large-scale sociotechnical networks, it is argued, are self-adjusting and self-correcting. Their coherence and rationality depend upon thousands of intelligent choices, bargains, and adjustments made at the various intersections of the organized system. Extended to the relationships throughout society, this process brings harmony and intelligence to the mass of fragmented, narrowly focused parts.

The most interesting statement of this position is found in Charles Lindblom's philosophy of "disjointed incrementalism."[19] Lindblom criticizes the "rational-comprehensive model" of decision making developed by Herbert Simon and others. He observes that the model's quest for an all-encompassing overview "assumes intellectual capacities and sources of information that men simply do not possess."[20] Societies and organizations are much too large and complex to allow decision makers a synoptic vision of all factors relevant to policy choices. But following the incrementalist "strategy," this makes little difference.

Under the "strategy," which Lindblom believes organizations actually do and certainly ought to follow, all moves are small ones. Changes are made step by step with careful attention to existing policies and conditions. As he puts it, "Only those policies are considered whose known or expected consequences differ incrementally from the status quo."[21] All that a decision maker needs to understand is the specific context of the particular choice before him. Knowledge of the activities of the whole system is irrelevant except as it bears directly on the incremental step at hand. The search for any broader perspective on the situation should be considered a waste of time and energy.

The Lindblomian scheme of things has implications far beyond the concerns of organization theory. Taken as a vision of social reality, disjointed incrementalism explains the relationship of knowledge to political action in the fragmented, complex mass of modern society. Most importantly, it tells us why a dearth of knowledge about the values of the whole of society and concern for those values causes no problem. Taken at its word, incrementalism offers an elaborate justification for ignoring the concept of "the public" once and for all.

In "The Science of Muddling Through" Lindblom tests his view from the standpoint of government agencies. "Suppose that each value neglected by one policy-making agency were a major concern of at least one other agency. In that case, a helpful division of labor would be achieved, and no agency need find its task beyond its capacities."[22] "The virtue of such a hypothetical division of labor is that every important interest or value has its watchdog."[23] Of course this is more than hypothetical. Political scientists, Lindblom among them, argue that this is how things actually work. Pluralism in the polity, as expressed by organized interests, corresponds to pluralism in the government bureaucracy. "Without claiming that every interest has a sufficiently powerful watchdog, it can be argued that our system often can assure a more comprehensive regard for the values of the whole society than any attempt at intellectual comprehensiveness."[24]

In the end, Professor Lindblom's theory isolates the unintelligibility of society and accompanying bafflement of the intellect as the "intelligence of democracy" itself.[25] The apparent strength of incrementalist theory, however, is also the source of an interesting puzzle. The model describes a massive, fragmented complex political system that, admittedly, no one fully comprehends. Its promise is that if each of the fragments comprehends and takes care of its own little sphere, the whole will run smoothly and with the maximum possible good. But what is it about disjointed incremental action that produces a society of harmonious relationships? How is it that such a beneficent result is achieved through the work of groups and individuals with little knowledge of or concern for the good of each other or the good of society as a whole? Is there a deus ex machina that intervenes to bring this wonderful result?

At this point the idea of autonomous technology, seen as an aspect of the complexity and unintelligibility of man's sociotechnical environment, merges with the classic problem of the "invisible hand." In Adam Smith's famous passage, the individual "intends only his own security; and by directing that industry in such a manner as its produce may be of the greatest value, he intends only his own gain, and he is in this . . . led by an invisible hand to promote an end which was no part of his intention. Nor is it always the worse for society that it was no part of it. By pursuing his own interest he frequently promotes that of society more effectually than when he really intends to promote it."[26]

Lindblom's contribution to this classic puzzle is to show that adjustments take place through bargaining and to argue that the coherency of the whole is insured so long as change occurs in moderate increments. "Societies, it goes without saying," he says, "are complex structures that can avoid dissolution or intolerable dislocation only by meeting certain preconditions, among them that certain kinds of change are admissible only if they occur slowly."[27]

But what if the invisible hand, even in the up-to-date Lindblomian version, should lose its touch? Is there a level of complexity or rapidity of change (or the two combined) in which the beneficence of the aggregate process is no longer insured? Indeed, there are times when the technological society appears to go incrementally mad. Devoid of any clear vision of its purposes or adequate science of its own workings, the complex self-adjusting mechanism rushes onward following some beneficial paths and others that are manifestly destructive. Looking, for example, at the modern city, one finds that the invisible hand apparently has a mischievous counterpart that brings confusion, chaos, and social ills on a colossal scale. At the center of such problems are technical systems of various kinds, always intended to "improve things," which in their presence in the complex social fabric generate as many dysfunctions as benefits. Thus, the question becomes: Which of two invisible hands—the benevolent or malevolent—has the firmest grasp? Which hand is more likely to thrive in a situation of increasing confusion?

In situations of this kind, large, complicated technologies have a paradoxical role. They may indeed help to "solve" one or another of the problems society encounters. At the same time, each addition to the technological aggregate contributes to the process of helpless drift, which itself defies solution. This is the condition Paul Goodman called "the metaphysical emergency of Modern Times: feeling powerless in immense social organizations; desperately relying on technological means to solve problems caused by previous technological means; when urban areas are technically and fiscally unworkable, extrapolating and planning for their future growth. Then, 'Nothing can be done.' "[28] As long as our sociotechnical systems continue to produce satisfactory conditions, as long as they function without a major hitch, complexity appears to pose no problem. But when things begin to go awry, when the technological order begins to generate surprising and undesirable outcomes, then the consequences of the complexity of the modern artifacts of civilization loom large. And one scarcely knows where to begin.

There is, obviously, much more that could be said about these matters. If one were to pursue the question, one would certainly want to inquire into the kinds of responses men and women have made to a world of artificial phenomena of such mass, power, and complication. What has been the response of intellectuals, those who take it upon themselves to develop means of making the major structures of culture intelligible and meaningful? What, on the other hand, has been the response of ordinary people as they move through the vast networks trying to find their way about?

The quest for satisfactory answers would take us far beyond the boundaries of the present work. I think, however, that a brief version may look something like this. Both the intellectual and the ordinary person in technological society are able to find satisfaction in images and interpretations that in some ways represent the world in which they exist. But the means of interpretation are themselves increasingly artificial, narrowly focused, and designed for particular productive effects. The ability to provide an intelligible account of the world on a broad scale is simply beyond their power. For the intellectual this means that an astounding proliferation of models, theories, and world views invades the world of thought, each of them highly specialized, tentative, and self-consciously artificial. Such "tools of inquiry" are understood to be heuristic, suggestive of fruitful interpretations, or predictive of some relatively small range of variables. But since there are frequent model changes and incessant retoolings, any initial confidence in any one mode of interpretation diminishes rapidly. The knowledge that "this too is an artifact" requires that intellect must keep moving. It never comes to rest, never finds a permanent home.

In much the same way, the ordinary citizen must rely on signals transmitted by the mass media. He immerses himself in the metaworld of shows, extravaganzas, commercials, news, and televised sports events and allows them to represent a larger world that he cannot experience firsthand.[29] He knows better than to expect truth from the endless stream of programs. Instead he seeks satisfaction, titillation, and a minimal level of real information. He rests content in the belief that if anything does happen, there will be a televised special on it. Thus, the shape of history comes to be seen as a sequence of disconnected specials, which temporarily "interrupt our regularly scheduled programming." Along with other media "events," which are known to be pure contrivance, the "news" itself takes on the aura of contrivance. Human beings do make their own history. Yes, but they do so in the same way that they make a situation comedy or the TV Game of the Week.

The Loss of Agency in Technological Systems

In summary, the position developed here suggests that members of the technological society actually know less and less about the fundamental structures and processes sustaining them. The gap between the realities of the world and the pictures individuals have of that world grows ever greater. For this reason, the possibility of directing technological systems toward clearly preceived, consciously chosen, widely shared aims becomes an increasingly dubious matter. Most persons are caught between the narrowness of their everyday concerns and a bedazzlement at the works of civilization. Beyond a certain point they simply do not know or care about things happening in their surroundings. With the overload of information so monumental, possibilities once crucial to citizenship are neutralized. Active participation is replaced by a haphazard monitoring. Thus, the technological order and its major subcomponents, through paths already traced, are free to take on a character of their own, which determines their destination. What one finds therefore, are highly developed systems of control, which are themselves beyond intelligence, beyond control, propelled toward goals that can be understood only by studying those systems' own peculiar inertia.[30]

The most important consequence of this situation is that, in a fundamental way, the whole society runs off track. The idea that civilized life consists of a fully conscious, intelligent, self-determining populace making informed choices about ends and means and taking action on that basis is revealed as a pathetic fantasy.

But there are other consequences to be mentioned if this aspect of the discussion is to make even a scant beginning. Under the conditions described here one finds interesting problems of action in complex technological aggregates. On occasion a particular system generates an unexpected event whose origins and cause are a puzzle. Mistakes, errors, and extraordinary lapses of control occur which simply should not have happened. Deeds and misdeeds are done which seem to defy any reasonable account. Certainly such occurrences are the exception. But in the twentieth century they have taken place often enough to indicate that the foundations of technological society are less reliable than some had hoped.

In the following cases, taken from the history of modern warfare, I want briefly to illustrate some difficulties of perception, control, and responsibility which sometimes afflict complex technological systems.

A fascinating document from the *Pentagon Papers* supplies our first example. The "Vietnam Bombing Evaluation by the Institute for Defense Analysis" tells of a long period of time in which one large-scale military system did not effectively perceive the circumstances of its own action. The time was the summer of 1966 when United States forces were conducting intensive bombing raids against the North Vietnamese. After months of study the IDA experts delivered a stern judgment: "As of July 1966 the U.S. bombing of North Vietnam (NVN) had no measurable direct effect on Hanoi's ability to mount and support military operations in the South at the current level."[31] The report gives reasons why this was true. Among its findings is this: "The bombing clearly strengthened popular support of the regime by engendering patriotic and nationalistic enthusiasm to resist the attacks."[32]

But the specifics are less interesting to us now than the IDA evaluation of the broader circumstances in which bombing policies were made and carried out. The report notes that "initial plans and assessments for ROLLING THUNDER program clearly tended to overestimate the persuasive and disruptive effects of the U.S. air strikes and correspondingly, to underestimate the tenacity and recuperative capabilities of the North Vietnamese."[33] Even more important, however, was the fact that, once underway, there was no effective means of comprehending the futility of the policy. According to the plan, the bombing would disrupt the supply of North Vietnamese resources and effectively break the will of the enemy to fight. The report observes: "These two sets of interrelationships are assumed in military planning, but it is not clear that they are systematically addressed in current intelligence estimates and assessments. Instead, the tendency is to encapsulate the bombing of NVN as one set of operations and the war in the South as another set of operations, and to evaluate each separately; and to tabulate and describe data on the physical, economic, and military effects of the bombing but not to address specifically the relationships between such effects and the data relating to the ability and will of the DRV (Democratic Republic of Vietnan) to continue its support of the war in the South."[34]

In other words, the intelligence estimates were good at counting the number of bridges, roads, and railroad tracks knocked out. But this information had no direct bearing on the plans and policies in question.

The IDA's conclusion is eloquent in its anguish over what the Pentagon had been doing:

The fragmented nature of current analyses and the lack of adequate methodology for assessing the net effects of a given set of military operations leaves a major gap between the quantifiable data on bomb damage effects, on the one hand, and policy judgments about the feasibility of achieving a given set of objectives, on the other. Bridging this gap still requires the exercise of broad political-military judgments that cannot be supported or rejected on the basis of systematic intelligence indicators. It must be concluded, therefore, that there is currently no adequate basis for predicting the levels of U.S. military effort that would be required to achieve the stated objectives—indeed, there is no firm basis for determining if there is any feasible level of effort that would achieve these objectives.[35]

Not wanting to be purely negative, the report closes with a few suggestions. It strongly recommends the building of a multisystem barrier across the demilitarized zone, employing the latest in military and communications hardware. "Weapons and sensors which can make a much more effective barrier, only some of which are now under development, are not likely to be available in less than 18 months to 2 years. Even these it must be expected, will eventually be overcome by the North Vietnamese, so that further improvements in weaponry will be necessary. Thus we envisage a dynamic 'battle of the barrier,' in which the barrier is repeatedly improved and strengthened by the introduction of new components, and which will hopefully permit us to keep the North Vietnamese off balance by continually posing new problems for them."[36]

In the grand style of twentieth-century technomania, the IDA analysts propose that the Pentagon move from round one to round two in the technology of destruction. Is it possible, however, that the new system, the electronic battlefield, would produce just as great or even greater dissociation from reality? Harold L. Wilensky has argued that "information pathologies" are endemic in large-scale organizations.[37] "In all complex social systems, hierarchy, specialization and centralization are major sources of distortion and blockage of intelligence."[38] While some of these difficulties can with cost be minimized, the source is never fully eliminated. "Intelligence failures are rooted in structural problems that cannot be fully solved; they express universal dilemmas of organizational life."[39] What bizarre foulups and errors will result from this is a less interesting question if your own life is not at stake. Cross-eyed, out of focus, unsure of the exact location of its own mighty appendages, the machine prepares to smash its next victims.[40]

From an earlier period of time comes an interesting incident, which would now be classified as a problem of "command and control." Are large, complex technological systems always amenable to guidance, even by those in the most obvious and powerful positions of control?

On August 1, 1914, Europe balanced precariously on the brink of holocaust. The Russians had failed to respond to Germany's latest ultimatum. In response, Kaiser Wilhelm declared a general mobilization, which sent the Schliefen plan into action. The plan for Germany's two-front war had been in various stages of preparation since 1892. It called for a swift and decisive strike against the western theater of battle. Barbara Tuchman describes the situation as follows: "Once the mobilization button was pushed, the whole vast machinery for calling up, equipping, and transporting two million men began turning automatically. . . . **One army corps alone—out of the total of 40 in the German forces** —required 170 railway cars for officers, 965 for infantry, 2,960 for cavalry, 1,915 for artillery and supply wagons, 6,010 in all, grouped in 140 trains and an equal number again for their supplies. From the moment the order was given, everything was to move at fixed times according to a schedule precise down to the number of train axles that would pass over a given bridge within a given time."[41]

Later in the day the kaiser began to have misgivings about the whole matter. He had received word from his ambassador in London that the British might be willing to intervene to keep France neutral in the coming conflict. This would give Germany a one-front war or possibly no war at all. The kaiser immediately summoned his chief of staff, General

Helmuth von Moltke, and ordered that the forward thrust of the Schliefen plan be stopped. The general's answer was surprising.

"Your Majesty," Moltke said to him now, "it cannot be done. The deployment of millions cannot be improvised. If Your Majesty insists on leading the whole army to the East it will not be an army ready for battle but a disorganized mob of armed men with no arrangements for supply. Those arrangements took a whole year of intricate labor to complete"—and Moltke closed upon that rigid phrase, the basis for every major German mistake, the phrase that launched the invasion of Belgium and the submarine war against the United States, the inevitable phrase when military plans dictate policy—"and once settled, it cannot be altered."[42]

Tuchman points to historical evidence which indicates that, in fact, the Schliefen plan could have been halted. But this argument was of no comfort to Kaiser Wilhelm on that summer's day. What he experienced was a vast system of warfare triggered and then unstoppable even by his imperial command. His knowledge included only those steps required to start the complex mechanism rolling. Beyond that his understanding was not sufficient to allow him effective action. Once he had given the sign, the deed was for all intents and purposes irreversible.

True, such occurrences are fairly rare in the twentieth century. But when they do occur their ramifications are completely lethal. The Vietnam war gives us at least one example of a diplomatic peace feeler that was destroyed by an air strike programmed weeks in advance which no one remembered to cancel.[43] And it was not so long ago, the reader will recall, that steps were taken to diminish the danger that the complex workings of military systems would set off an accidental nuclear war—a possibility that was for a time very real. The incomprehensibility, inflexibility, and tendency toward destructive inertia of such systems pose a continuing danger to the human populations of the earth, including those segments in positions of ostensible control. That such dilemmas are both in principle and in technique usually solvable offers little comfort, for it may occur that the last unseen, unremedied lapse of control will be the one that finishes the story.

A final group of illustrations bears on the question of responsibility in large-scale systems. The twentieth century has given a peculiar turn to the context in which deeds and misdeeds take place. One finds that it is sometimes very difficult to locate praise or blame for events that occur within massive aggregates of men and machinery. Upon inspection one finds no person able or willing to say: "I did this thing. I knew what I was doing. I will accept the consequences."

We have all had experiences of this kind on a trivial level. Because of a mistake in bureaucratic paperwork, a document important to one's activities is misplaced. An error in a computerized bank statement or credit card account stirs up temporary problems in one's finances. In most such cases we simply say "Something must have gone wrong somewhere along the line." In a few days or weeks the disruption is usually repaired. But the system at hand is too vast and the error too insignificant to make the search for its specific origin worthwhile.

On a higher level of size and complexity, however, there are often important problems surrounding the origin of events. A late 1960s study of the workings of the Department of Defense by a presidential Blue Ribbon Panel, reported extreme frustration in this regard. Within the complicated structure of the Pentagon, the panel was often unable to find where important decisions were made and who was subject to praise or blame.[44]

Circumstances of this sort enter the arena of history with stories like that of Adolf Eichmann. At his trial in Jerusalem in 1961, Eichmann's defense argued that his guilt was strongly mitigated by the fact that he was only a "small cog" in a vast system.[45] He had been in charge of a set of railroad lines. His job was to see that the trains ran efficiently. The larger context into which this transportation fit was not his concern. His role in the bureaucracy and his specific orders were his only interest. In Hannah Arendt's words: "As for the base motives, he was perfectly sure that he was not what he called an innerer Schweinehund, a dirty bastard in the depths of his heart; and as for his conscience, he remembered perfectly well that he would have had a bad conscience only if he had not done what he had been ordered to do—to ship millions of men, women, and children to their death with great zeal and the most meticulous care."[46] Like so many other systems employees of our age, Eichmann had a job to do and he did it.

To most of the world, nevertheless, Eichmann's defense did not wash. His excuse was a lame one, particularly when compared with the magnitude of his crime. But at a later date, the trial of Lieutenant William Calley for the massacre of civilians at My Lai, the same kind of plea was introduced. Calley's case soon became an American cause célèbre. His defense argued that he could not have been responsible for the killings at the village since he was a mere cipher in the U.S. Army's vast mechanism. He was, in effect, too close to the deed to be at fault. Calley, it was suggested, was being used as a scapegoat for those higher up in authority. But as one looked upward, one found that the argument was one of diminished responsibility at that level as well. Those in positions of near or distant command *did not know* (or said they did not know) what was happening. Since they could not control everything that occurred out in the field, it followed that they could not reasonably be held to blame for the events that took place.

The logic here, described in full, is a wonder to behold. Both proximity and distance count as excuses. The closer you are, the more innocent; the farther away you are, the more innocent. It is a magnificent arrangement in which everyone is safe except the victims. In a system of this kind the very notion of a "deed" seems to evaporate. The concept of responsibility becomes as slippery as a squid in a fish market bin. Difficulties in tracing origins—trivial in the case of one's bank statement—take on monumental proportions when the issue, as it was with Eichmann and Calley, becomes genocide.

Some possible ramifications of this moral context are now at issue in a variety of sophisticated technology mentioned earlier—the electronic battlefield. In a report edited by Raphael Littauer and Norman Uphoff, *The Air War in Indochina*, is a description of a complex system which apparently increases both the possibility of lethal error and the actors' sense of remoteness from any concrete responsibility.

When friend and foe are intermingled, how can electronic sensing and controlling devices discriminate between them? This remains a basic problem under any conditions. A seismic detector cannot tell the difference between a truck full of arms and a school bus; a urine sniffer cannot tell a military shelter from a woodcutter's shack. The further the U.S. goes down the road to automation, and the greater its capital investment becomes relative to its investment in manpower, the more deeply will it become committed to this blind form of warfare.

The human operator . . . is terribly remote from the consequences of his actions; he is most likely to be sitting in an air-conditioned trailer, hundreds of miles from the area of battle; from there he assesses "target signatures," evaluates ambiguities in the various sensor systems, collates their reports, and determines the tactical necessity for various forms of action which are then implemented automatically. For him, the radar blip and flashing lights no more represent human beings than the tokens in a board type war game. War and war games become much the same.[47]

One primary value of any system of responsibility is that it provides an element of restraint. Excesses which might conceivably occur are limited at their source, the consciousness and acts of individuals. To the best of my knowledge, there have been no rigorous studies of how automated warfare has affected the psychology of military action. But the Littauer and Uphoff study gives some early sounding. It reports the testimony of a retired general who participated in war games at the National War College and who noticed a regular pattern of responses among his fellows. "Those officers whose weapons systems delivered death remotely were much more willing to call awesome amounts of firepower into play."[48]

In classical ethics a person is excused from blame for a misdeed if sufficient extenuating circumstances can be shown to exist. For example, an automobile driver who injures a pedestrian may claim that his responsibility is mitigated because he swerved to avoid a speeding automobile. What is interesting about the new ethical context offered by highly complex systems is that their very architecture constitutes vast webs of extenuating circumstances. Seemingly valid excuses can be manufactured wholesale for anyone situated in the network. Thus, the very notion of moral agency begins to dissolve.

Recognition of difficulties in placing praise or blame for the behavior of complex systems is sometimes the cause for great delight among some members of the technical intelligentsia. Computer scientists have long hoped that truly autonomous computer programs—programs that perform unpredictably, creatively, and beyond the comprehension of their makers—would be a sign that work in artificial intelligence had made substantial progress. Commenting on the views of A. L. Samuel on this matter, Marvin Minsky writes, "His argument, based on the fact that reliable computers do only that which they are instructed to do, has a basic flaw; it does not follow that the programmer therefore has full knowledge of (and therefore full responsibility and credit for) what will ensue. For certainly the programmer may set up an evolutionary system whose limitations are for him unclear and possibly incomprehensible."[49] Minsky may believe that the products of such incomprehensible programs will be uniformly positive, that is, that programs that behave in ways its makers could not predict will achieve marvels of performance in, for example, games of checkers. But what is to prevent possible detrimental effects following from such work? What is to stop an incomprehensible program from developing a lethal edge? Certainly not the programmer. In the very nature of the situation, he is, according to Minsky, not truly entitled to "full credit" for his program's achievements, and, what is more to the present point, he cannot be charged with "full responsibility" for his program's possibly devastating effects.

The problems for moral agency created by the complexity of technical systems cast new light on contemporary calls for more ethically aware scientists and engineers. According to a very common and laudable view, part of the education of persons learning advanced scientific skills ought to be a full comprehension of the social implications of their work. Enlightened professionals should have a solid grasp of ethics relevant to their activities. But, one can ask, what good will it do to nourish this moral sensibility and then place the individual in an organizational situation that mocks the very idea of responsible conduct? To pretend that the whole matter can be settled in the quiet reflections of one's soul while disregarding the context in which the most powerful opportunities for action are made available is a fundamental misunderstanding of the quality genuine responsibility must have.

Already, in fact, conditions of unintelligibility and extenuation of blame have set the stage for a novel doctrine of organizational cynicism —"plausible deniability." Here, even those aware of unfortunate or unsavory occurrences can arrange (before the fact) that their complex environment prevents crucial information from reaching them and "truthfully" claim later that they simply did not know. As revelations from Watergate and the dealings of America's "intelligence" community suggest, the idea of plausible deniability vastly increases the possibility of tolerating evil until it is too late.

In the *Eclipse of Reason* Max Horkheimer observes that "as material productions and social organization grow more complicated and reified, recognition of means as such becomes increasingly difficult, since they assume the appearance of autonomous entities."[50] This states the case very well. My only quarrel with Horkheimer's formulation is its emphasis on "appearance." There are times in which the difference between the distressing appearance and untenable reality of these autonomous entities blurs totally. At such times it is the faith that "man controls technology," rather than the contrary view, which looms as an irrational belief.

Chapter 8
Frankenstein's Problem

Our inquiry began with the simple recognition that ideas and images of technology-out-of-control have been a persistent obsession in modern thought. Rather than dismiss this notion out of hand, I asked the reader to think through some ways in which the idea could be given reasonable form. The hope was that such an enterprise could help us reexamine and revise our conceptions about the place of technology in the world. In offering this perspective, I have tried to indicate that many of our present conceptions about technics are highly questionable, misleading, and sometimes positively destructive. I have also tried to lay some of the early groundwork for a new philosophy of technology, one that begins in criticism of existing forms but aspires to the eventual articulation of genuine, practical alternatives.

A possible objection to the notions I have developed here is that they are altogether Frankensteinian. Some may suppose that in choosing this approach, the inquiry enters into an old and discredited myth about the age of science and technology. It is easy, indeed, for the imagination to get carried away with images of man-made monstrosities. And it is not difficult to get snagged on the linguistic hook which allows us to talk about inanimate objects with transitive verbs, as if they were alive. Some may even conclude that such traps fulfill a certain need and that autonomous technology is nothing more than an irrational construct, a psychological projection, in the minds of persons who, for whatever reason, cannot cope with the realities of the world in which they live.

To doubts of this kind I would reply: Does the point refer to the book or the motion picture? What, after all, is Frankenstein's problem? What exactly is at stake in the notorious archetypical inventor's relationship to his creation? Unfortunately, the answers to this question now commonly derive not from Mary Wollstonecraft Shelley's remarkable novel but from an endless stream of third-rate monster films whose makers give no indication of having read or understood the original work. Here is a case in which the book is truly superior to the movie. The Hollywood retelling fails to notice the novel's subtitle, "A Modern Prometheus," much less probe its meaning. The filmmakers totally ignore the essence of the story written by a nineteen-year-old woman, daughter of the radical political theorist William Godwin and one of the earliest of militant feminists, Mary Wollstonecraft. As a consequence, no justice is done to a work that it seems to me is still the closest thing we have to a definitive modern parable about mankind's ambiguous relationship to technological creation and power.

In the familiar Hollywood version, the story goes something as follows. A brilliant but deranged young scientist constructs a hideous creature from human parts stolen from graveyards. On a stormy evening in the dead of winter the doctor brings his creature to life and celebrates his triumph. But there is a flaw in the works. The doctor's demented assistant, Igor, has mistakenly stolen a criminal brain for the artificial man. When the monster awakes, he tears up the laboratory, smashes Doctor Frankenstein, and escapes into the countryside killing people right and left. The doctor is horrified at this development and tries to recapture the deformed beast. But before he can do so, the local townspeople chase down the monster and exterminate him. This ending is, of course, variable and never certain, lending itself to the needs for a plot—over a forty-year period—for Boris Karloff, Bela Lugosi, Lon Chaney, Jr., Peter Cushing, and Mel Brooks.

The fact of the matter is that the film scenarios have virtually nothing to do with *Frankenstein* the novel. In the original there is no crazed assistant, no criminal brain mistakenly transplanted, no violent rampage of random terror, no final extermination of the creature to bring safety and reassurance (although there is mention of a graveyard theft). In the place of such trash, the book contains a story offering an interesting treatment of the themes of creation, responsibility, neglect, and the ensuing consequences. Let us see what Mary Shelley's gothic tale actually has to say.

From the time of his youth the young Genevan, Victor Frankenstein, was fascinated by the causes of natural phenomena. "The world," he tells us, "was to me a secret which I desired to divine. Curiosity, earnest research to learn the hidden laws of nature, gladness akin to rapture, as they were unfolded to me, are amongst the earliest sensations I can remember."[1] As he reaches maturity, his first response to this lingering obsession is to probe alchemy and the occult, the texts of Albertus Magnus and Paracelsus, in quest of the philosopher's stone. But realizing the futility of this research, he soon turns to the new science of Bacon and Newton. He hears a professor tell how the modern masters are superior to the ancients since they "penetrate into the recesses of nature and show how she works in her hiding places,"[2] a distinctly Baconian notion of what is involved. Following the principles of mathematics and natural philosophy, he eventually comes upon "the cause of the generation of life; nay more, I became myself capable of bestowing animation upon lifeless matter."[3]

To this point the story sounds very much like the one we all think we know. But from here on the novel takes some surprising turns. One evening Victor Frankenstein does bring his artificial man to life. He sees it open its eyes and begin to breathe. But instead of celebrating his victory over the powers of nature, he is seized by a rash of misgivings. "Now that I had finished, the beauty of the dream vanished, and breathless horror and disgust filled my heart. Unable to endure the aspect of the being I had created, I rushed out of the room and continued a long time traversing my bedchamber, unable to compose my mind to sleep."[4] And what about the newborn "human" back in the laboratory? He is left to his own devices trying to figure out what in the world has happened to him. Quietly he walks to Victor's bedroom, draws back the bed curtain, smiles, and tries to speak. But Victor, in the throes of a crisis of nerve, is still not ready to accept the life that he brought into existence and simply panics. "He might have spoken, but I did not hear; one hand was stretched out, seemingly to detain me, but I escaped and rushed downstairs. I took refuge in the courtyard belonging to the house which I inhabited, where I remained during the rest of the night, walking up and down in the greatest agitation, listening attentively, catching and fearing each sound as if it were to announce the approach of the daemonical corpse to which I had so miserably given life."[5]

Thus, it is Frankenstein himself who flees the laboratory, *not* his benighted creation. The next morning Victor leaves the house altogether and goes to a nearby town to tell his troubles to an old friend. This is very clearly a flight from responsibility, for the creature is still alive, still benign, left with nowhere to go, and, more important, stranded with no introduction to the world in which he must live. Victor's protestations of misery, remorse, and horror at the results of his work sound particularly feeble. It is clear, for example, that the monstrosity of his creation is in the first instance less a matter of its physical appearance than of Frankenstein's terror at his own success. He is haunted henceforth not by the creature itself but by the vision of it in his imagination. He does not return to his laboratory and makes no arrangements of any kind to look after his work of artifice. The next encounter between the father and his technological son comes more than two years later.

An important feature of *Frankenstein,* the feature of the book that makes it useful for our purposes, is that the artificial being is able to explain his own position. Fully a third of the text is either "written" by his hand or spoken by him in dialogue with his maker. After his abandonment in the laboratory, the creature leaves the place and enters the world to make his way. Eventually, he takes up residence in a forest near a cottage inhabited by a Swiss family. He eavesdrops on them, notices how they use words, and after a while masters language himself. Stumbling upon a collection of books, he teaches himself to read and soon finishes off *Paradise Lost,* Plutarch's *Lives,* and the *Sorrows of Young Werther.* Later he examines the coat he had carried with him from the laboratory and finds Frankenstein's diary describing the circumstances of the experiment and giving the true identity of his maker. When the creature finally meets Victor on an icy slope in the Alps, he is ready to state an eloquent case. Autonomous technology personified finds its voice and speaks. The argument presented emphasizes the perils of an unfinished, imperfect creation, cites the continuing obligations of the creator, and describes the consequences of further insensitivity and neglect.

"I am thy creature, and I will be even mild and docile to my natural lord and king if thou wilt also perform thy part, that which thou owest me."[6]

"You propose to kill me. How dare you sport thus with life? Do your duty towards me, and I will do mine towards you and the rest of mankind. If you will comply with my conditions I will leave them and you at peace; but if you refuse, I will glut the maw of death, until it be satiated with the blood of your remaining friends."[7]

The monster explains that his first preference is to be made part of the human community. Frankenstein was wrong to release him into the world with no provision for his role or influence in the presence of normal men. Already his attempts to find a home have had disastrous results. He introduced himself to the Swiss family, only to find them terrified at his grotesque appearance. On another occasion he unintentionally caused the death of a young boy. He now asks Frankenstein to recognize that the invention of something powerful and novel is not enough. Thought and care must be given to its place in the sphere of human relationships. But Frankenstein is still too thick and self-interested to comprehend the message. "Abhorred monster! Fiend that thou art! . . . Begone! I will not hear you. There can be no community between you and me; we are enemies. Begone, or let us try our strength in a fight, in which one must fall."[8]

Despite this stream of invective, the creature continues to reason with Victor. It soon becomes apparent that he is, if anything, the more "human" of the two and the man with the better case. At the same time, he leaves no doubt that he means business. If no accommodation is made to his needs, he will take revenge. After a while Victor begins to yield to the logic of the monster's argument. "For the first time," he admits, "I felt what the duties of a creator towards his creature were, and that I ought to render him happy before I complained of his wickedness."[9] The two are able to agree that it is probably too late for the nameless "wretch" to enter human society, and they arrive at a compromise solution: Frankenstein will return to the laboratory and build a companion, a female, for his original masterpiece. "It is true, we shall be monsters, cut off from all the world; but on that account we shall be more attached to one another."[10] The problems caused by technology are to find a technological cure.

Of course, the scheme does not work. After a long period of procrastination, Victor sets to work on the second model of his invention, but in the middle of his labors he remembers a pertinent fact. The first creature "had sworn to quit the neighborhood of man and hide himself in deserts, but she had not; and she, who in all probability was to become a thinking and reasoning animal, might refuse to comply with a compact made before her creation."[11] The artificial female would have a life of her own. What was to guarantee that she would not make demands and extract the consequences if the demands were not properly met? Then an even more disquieting thought strikes Victor. What if the two mate and have children? "A race of evils would be propagated upon the earth who might make the very existence of the species of man a condition precarious and full of terror."[12] "I shuddered to think that future ages might curse me as their pest, whose selfishness had not hesitated to buy its own peace at the price, perhaps, of the existence of the whole human race."[13] Recognizing what he believes to be a heroic responsibility, Victor commits an act of violence. With the first creature looking on, he tears the unfinished female artifact to pieces.

From this point the story moves toward a melodramatic conclusion befitting a gothic novel. The creature reminds Victor, "You are my creator, but I am your master," and then vows, *"I will be with you on your wedding night."*[14] He makes good his promise and eventually kills Victor's young bride Elizabeth. Frankenstein then sets out to find and destroy his creature, but after a long period without success succumbs to illness on a ship at sea. In the final scene the creature delivers a soliloquy over Victor's coffin and then floats on an ice raft, announcing that he will commit suicide by cremating himself on a funeral pyre.

In recent years it has become fashionable to take *Frankenstein* seriously. The book frequently appears as the subject of elaborate psychosexual analyses, which seize upon some colorful episodes in the relationships of Mary Shelley, her famous mother and father, as well as Percy Bysshe Shelley and friend of the family, Lord Byron.[15] There is no doubt some truth to these interpretations. The book abounds with pointed references to problems of sexual identity, child-parent conflicts, and love-death obsessions. But there is also adequate evidence that in writing her story Shelley was also interested in the possibilities of science and the problems of scientific invention. In her time, as in our own, it was not considered fantasy that the secrets of nature upon which life depends might be laid open to scrutiny and that this knowledge could be used to synthesize, in whole or in part, an artificial human being.[16] It is not unlikely in this regard that the book was meant as criticism of the Promethean ideals of her husband. Percy Shelley saw in the figure of Prometheus, rebel and life-giver, a perfect symbol to embody his faith in the perfectibility of man, the creative power of reason, and the possibility of a society made new through enlightened, radical reconstruction. His play *Prometheus Unbound* sees its hero released from the fetters imposed by the gods and freed for endless good works. In its preface Shelley explains that "Prometheus is, as it were, the type of the highest perfection of moral and intellectual nature, impelled by the purest and truest motives to the best and noblest ends."[17] To the charge that the poet himself has gotten carried away with "a passion for reforming the world," Shelley replies: "For my part I had rather be damned with Plato and Lord Bacon than go to Heaven with Paley and Malthus."[18] Plato and Lord Bacon?

Mary Shelley's novel, published at about the same time as her husband's play, may well have been an attempt to rediscover the tragic flaw in a vision from which Shelley hoped to eliminate any trace of tragedy. In the Baconian-Promethean side of her spouse's quest, the side that marveled at the powers that could come from the discovery and taming of nature's secrets, she found a hidden agenda for trouble. The best single statement of her view comes on the title page of the book, a quotation from Milton's *Paradise Lost:*

Did I request thee, Maker, from my clay
To mould me man? Did I solicit thee
From darkness to promote me?—

Suggested in these words is, it seems to me, the issue truly at stake in the whole of *Frankenstein:* the plight of things that have been created but not in a context of sufficient care. This problem captures the essence of the themes my inquiry has addressed.

Victor Frankenstein is a person who discovers, but refuses to ponder, the implications of his discovery. He is a man who creates something new in the world and then pours all of his energy into an effort to forget. His invention is incredibly powerful and represents a quantum jump in the performance capability of a certain kind of technology. Yet he sends it out into the world with no real concern for how best to include it in the human community. Victor embodies an artifact with a kind of life previously manifest only in human beings. He then looks on in surprise as it returns to him as an autonomous force, with a structure of its own, with demands upon which it insists absolutely. Provided with no plan for its existence, the technological creation enforces a plan upon its creator. Victor is baffled, fearful, and totally unable to discover a way to repair the disruptions caused by his half-completed, imperfect work. He never moves beyond the dream of progress, the thirst for power, or the unquestioned belief that the products of science and technology are an unqualified blessing for humankind. Although he is aware of the fact that there is something extraordinary at large in the world, it takes a disaster to convince him that the responsibility is his. Unfortunately, by the time he overcomes his passivity, the consequences of his deeds have become irreversible, and he finds himself totally helpless before an unchosen fate.

If the arguments we have examined have any validity at all, it is likely that Victor's problems have now become those of a whole culture. At the outset, the development of all technologies reflects the highest attributes of human intelligence, inventiveness, and concern. But beyond a certain point, the point at which the efficacy of the technology becomes evident, these qualities begin to have less and less influence upon the final outcome; intelligence, inventiveness, and concern effectively cease to have any real impact on the ways in which technology shapes the world.

It is at this point that a pervasive ignorance and refusal to know, irresponsibility, and blind faith characterize society's orientation toward the technical. Here it happens that men release powerful changes into the world with cavalier disregard for consequences; that they begin to "use" apparatus, technique, and organization with no attention to the ways in which these "tools" unexpectedly rearrange their lives; that they willingly submit the governance of their affairs to the expertise of others. It is here also that they begin to participate without second thought in megatechnical systems far beyond their comprehension or control; that they endlessly proliferate technological forms of life that isolate people from each other and cripple rather than enrich the human potential; that they stand idly by while vast technical systems reverse the reasonable relationship between means and ends. It is here above all that modern men come to accept an overwhelmingly passive response to everything technological. The maxim "What man has made he can also change" becomes increasingly scandalous.

Until very recently this adoption of an active image to mask the passive response seemed an entirely appropriate stance. The elementary tool-use conception of scientific technology, essentially unchanged since Francis Bacon, was universally accepted as an accurate model of all technical conduct. All one had to do was to see that the tools were in good hands. Reinforcing this view was a devout acceptance of the idea of progress, originally an ideal of improvement through enlightenment, the education of all mankind, and continuing scientific and technical advance. But eventually the technological side of the notion eclipsed the others. Progress came to be coterminous with the enlarging sphere of technological achievement. This was (and still is) widely understood to be a kind of ineluctable, self-generating process of increasing beneficence—autonomous change toward a desirable *telos*.

Beyond these dominant beliefs and attitudes, however, lies something even more fundamental, for there is a sense in which all technical activity contains an inherent tendency toward forgetfulness. Is not the point of all invention, technique, apparatus, and organization to have something and *have it over with?* One does not want to bother anymore with building, developing, or learning it again. One does not want to bother with its structure or the principles of its internal workings. One simply wants the technical thing to be present in its utility. The goods are to be obtained without having to understand the factory or the distribution network. Energy is to be utilized without understanding the myriad of connections that made its generation and delivery possible. Technology, then, allows us to ignore our own works. It is *license to forget.* In its sphere the truths of all important processes are encased, shut away, and removed from our concern. This more than anything else, I am convinced, is the true source of the colossal passivity in man's dealings with technical means.

I do not mean to overlook the fact that, on the whole, mankind has been well served in this relationship. The benefits in terms of health, mobility, material comfort, and the overcoming of the physical problems of production and communication are well known. That I have not recounted them frequently is not a sign that I have forgotten them. I live here too.

For the vast majority of persons, the simple, time-honored notions about technology are sufficient. There are still eloquent public spokesmen ready to explain the basic tenets at each suitable occasion. "When you talk about progress, about the new and the different, the possibilities are infinite. That's what is so fascinating and compelling about progress. The infinite possibilities. The potential for creating sights and sounds and feelings that have not yet been dreamed of, for achieving all that has yet to be achieved, for changing the world." "For it must be obvious to anyone with any sense of history and any awareness of human nature that there *will* be SST's. And Super SST's. And Super-Super SST's. Mankind is simply not going to sit back with the Boeing 747 and say 'This is as far as we go.' "[19]

For the many who embrace this faith, any criticism of technology is taken as vile heresy. Like Elijah defending Yahweh from the gods of

Jezebel and Ahab, the choice for them is strictly either/or, monotheism or not. Those who find problems in the technological content of this culture or who seriously suggest that different kinds of sociotechnical arrangements might be preferable are portrayed as absolute nay-sayers, pessimists, or, worse, crafty seducers luring innocent victims toward the brink of nameless dread.

There are, nevertheless, indications that the conversation is beginning to widen its boundaries. True, the Elijahs are still on Mount Carmel commanding piles of wood to catch fire (usually under some new aircraft or weapons system). Enthusiastic boosters and cheerleaders are still busy trying to obscure the fact that "progress" once meant something more than novel hardware and technique.[20] But other voices are beginning to speak. It is possible that a more vital, intelligent questioning is beginning to replace docile prejudice. Many now understand why it is necessary to think and act differently in the face of technological realities and to begin the search for new paths.

This book is intended as a contribution to the effort to reevaluate the circumstances of our involvement with technology. My aim has been to sketch in some detail problems I thought were underestimated or not sufficiently clear in other writings. The position of these perspectives is not, as the boosters may conclude, that technology is a monstrosity or an evil in and of itself. Instead, the view has been much like that of Mary Shelley's novel, that we are dealing with an unfinished creation, largely forgotten and uncared for, which is forced to make its own way in the world. This creation, like Victor's masterpiece, contains the precious stuff of human life. But in its present state it all too often returns to us as a bad dream—a grotesquely animated, autonomous force reflecting our own life, crippled, incomplete, and not fully in our control.

Is this a helpful conclusion to have drawn? Other than accounting for one recurrent problem in modern thought, what does it offer?

Very little. Very little, that is, unless those who build and maintain the technological order are willing to reconsider their work. Victor Frankenstein was blinded by two diametrically opposed beliefs: first, that he would produce an artifact of undeniable perfection and, later on, that his invention was a disaster about which nothing could be done. For those willing to go beyond both of these conclusions, the rest of the book offers a few more steps.

Technology as Legislation

Obviously there are a great many specific issues and approaches within the general range of questions we have encountered. The ecology movement, consumerism, future studies, the technology assessors, students of innovation and social change, and what remains of the "counterculture" all have something to say about the ways in which technology presents difficulties for the modern world. Since the reader is no doubt familiar with the debates now raging over these issues, I will not review the details. But in their orientations toward politics and their conceptions of how a better state of affairs might be achieved, the issue areas sort themselves into roughly two categories.

In the first domain, far and away the most prominent, the focus comes to rest on matters of risk and safeguard, cost and benefit, distribution, and the familiar interest-centered style of politics. Technology is seen as a cause of certain problematic effects. All of the questions raised in the present essay, for example, would be interpreted as "risks taken" and "prices paid" in the course of technological advance. Once this is appreciated, the important tasks become those of (1) accurate prediction and anticipation to alleviate risk, (2) adequate evaluation of the costs that are or might be incurred, (3) equitable distribution of the costs and risks so that one portion of the populace neither gains nor suffers excessively as compared to others, and (4) shrewd evaluation of the political realities bearing upon social decisions about technology.

Under this model the business of prediction is usually meted out to the natural and social sciences. Occassionally, some hope is raised that a new art or science—futurism or something of the sort—will be developed to improve the social capacity of foresight. The essential task is to devise more intelligent ways of viewing technological changes and their possible consequences in nature and society. Ideal here would be the ability to forecast the full range of significant consequences in advance. One would then have a precise way of assigning the risk of proceeding in one way rather than another.[21]

The matter of determining costs is left to orthodox economic analysis. In areas in which "negative externalities" are experienced as the result of technological practice, the loss can be given a dollar value. The price paid for the undesirable "side effects" can then be compared to the benefit gained. An exception to this mode of evaluation can be seen in some environmental and sociological arguments in which nondollar value costs are given some weight.[22] On the whole, however, considerations of cost follow the form Leibniz suggested for the solution of all rational disputes: "Let us calculate." Taking this approach one tends to ask questions of the sort: How much are you prepared to pay for pollution-free automobiles? What is the public prepared to tax itself for clean rivers? What are the trade-offs between having wilderness and open space as opposed to adequate roads and housing? Are the costs of jet airport noise enough to offset the advantage of having airports in the middle of town? Such questions are answered at the cash register, although the computer shows a great deal more style.

Once the risks have been assigned, the safeguards evaluated, and the costs calculated, one is then prepared to worry about distribution. Who will enjoy how much of the benefit? Who will bear the burden of the uncertainty or the price tag of the costs? Here is where normal politics—pressure groups, social and economic power, private and public interests, bargaining, and so forth—enters. We expect that those most aware, best supplied, and most active will manage to steer a larger proportion of the advantages of technological productivity their way while avoiding most of the disadvantages. But for those who have raised technology as a political problem under this conception, reforms are needed in this distributive process. Even persons who have no quarrel with the inequities of wealth and privilege in liberal society now step forth with the most trenchant criticisms of the ways in which technological "impacts" are distributed through the social system. A certain radicalism is smuggled in through the back door. The humble ideal of those who see things in this light is that risks and costs be allotted more equitably than in the past. Those who stand to gain from a particular innovation should be able to account for its consequences beforehand. They should also shoulder the major brunt of the costs of undesirable side effects. This in turn should eliminate some of the problems of gross irresponsibility in technological innovation and application of previous times. Since equalization and responsibility are to be induced through a new set of laws, regulations, penalties, and encouragements, the attention of this approach also aims at a better understanding of the facts of practical political decision making.

Most of the work with any true influence in the field of technology studies at present has its basis in this viewpoint. The ecology movement, Naderism, technology assessment, and public-interest science each have somewhat different substantive concerns, but their notions of politics and rational conduct all fit within this frame. There is little new in it. What one finds here is the utilitarian-pluralist model refined and aimed at new targets. In this form it is sufficiently young to offer spark to tired arguments, sufficiently critical of the status quo to seem almost risqué. But since it accepts the major premises and disposition of traditional liberal politics, it is entirely safe. The approach has already influenced major pieces of legislation in environmental policy and consumer protection. It promises to have a bright future in both the academic and the political realms, opening new vistas for "research," "policy analysis," and, of course, "consulting."[23]

On the whole, the questions I have emphasized here are not those now on the agendas of persons working in the first domain. But for those following this approach I have one more point to add. It is now commonly thought that what must be studied are not the technologies but their implementing and regulating systems. One must pay attention to various institutions and means of control—corporations, government agencies, public policies, laws, and so forth—to see how they influence the course our technologies follow. Fine. I would not deny that there are any number of factors that go into the original and continued employment of these technical ensembles. Obviously the "implementing" systems have a great deal to do with the eventual outcome. My question is, however, In what technological context do such systems themselves operate and what imperatives do they feel obliged to obey? In several ways I have tried to show that the hope for some "alternative implementation" is largely misguided. *That* one employs something at all far outweighs (and often obliterates) the matter of *how* one employs it. This is not sufficiently appreciated by those working within the utilitarian-pluralist framework. We may firmly believe that we are developing ways of regulating technology. But is it perhaps more likely that the effort will merely succeed in putting a more elegant administrative facade on old layers of reverse adapted rules, regulations, and practices?[24]

The second domain of issues is less easily defined, for it contains a collection of widely scattered views and spokesmen. At its center is the belief that technology is problematic not so much because it is the origin of certain undesirable side effects but rather because it enters into and becomes part of the fabric of human life and activity. The maladies technology brings—and this is not to say that it brings only maladies—derive from its tendency to structure and incorporate that which it touches. The problems of interest, therefore, do not arrive by chain reaction from some distant force. They are present and immediate, built into the everyday lives of individuals and institutions. Analyses that focus only upon risk/safeguard, cost/benefit, and distribution simply do not reveal problems of this sort. They require a much more extraordinary, deep-seeking response than the utilitarian-pluralist program can ever provide.

What, then, are the issues of this second domain? Some of the most basic of them are mirrored in our discussion of the theory of technological politics. This model represents the critical phase of a movement of thought, the attempt to do social and political analysis with technics as its primary focus. But these thoughts so far have given little care to matters of amelioration. In the present formulation of the theory, I have deliberately tried to avoid dealing in popular remedies. It is my experience that inquiries pointing to broad, easy solutions soon become cheap merchandise in the commercial or academic marketplace. They become props for the very thing criticized.[25]

For better or worse, however, most of the thinking in the second domain at present is highly specific, solution oriented, and programmatic. The school of humanist psychology, writers and activists of the counterculture, utopian and communal living experiments, the free schools, proponents of encounter groups and sensual reawakening, the hip catalogers, the peace movement, pioneers of radical software and new media, the founders and designers of alternative institutions, alternative architecture, and "appropriate" or "intermediate" technology—all of these have tackled the practical side of one or more of the issues raised in this essay.

Much of the work has begun with a sobering recognition of the psychological disorders associated with life in the technological society. The world of advanced technics is still one that makes excessive demands on human performance while offering shallow, incomplete rewards. The level of stress, repression, and psychological punishment that rational-productive systems extract from their human members is not matched by the opportunity for personal fulfillment. Men and women find their lives cut into parcels, spread out, and dissociated. While the neuroses generated are often found to be normal and productive in the sociotechnical network, there has been a strong revolt against the continuation of such sick virtues. Both professionals and amateurs in psychology have come together in a host of widely differing attempts to find the origins of these maladies and to eliminate them.

Other enterprises of this kind have their roots in a pervasive sense of personal, social, and political powerlessness. Confronted with the major forces and institutions that determine the quality of life, many persons have begun to notice that they have little real voice in most important arrangements affecting their activities. Their intelligent, creative participation is neither necessary nor expected. Even those who consider themselves "well served" have cause to wonder at decisions, policies, and programs affecting them directly, over which they exercise no effective influence. In the normal state of affairs, one must simply join the "consensus." One consents to a myriad of choices made, things built, procedures followed, services rendered, in much the same way that one consents to let the eucalyptus trees continue growing in Australia. There are some, however, who have begun to question this submissive, compliant way of life. In a select few areas, some people have attempted to reclaim influence over activities they had previously let slip from their grasp. The free schools, food conspiracies and organic food stores, new arts and crafts movement, urban and rural communes, and experiments in alternative technology have all—in the beginning at least—pointed in this direction. With mixed success they have sought to overcome the powerlessness that comes from meting out the responsibility for one's daily existence to remote large-scale systems.

A closely related set of projects stems from an awareness of the ways organized institutions in society tend to frustrate rather than serve human needs. The scandal of productivity has reached astounding proportions. More and more is expended on the useless, demeaning commodities idealized in the consumer ethos (for example, vaginal deodorants), while basic social and personal needs for health, shelter, nutrition, and education fall into neglect. The working structures of social institutions that provide goods and services seem themselves badly designed. Rather than elicit the best qualities of the persons they employ or serve, they systematically evoke the smallest, the least creative, least trusting, least loving, and least lovable traits in everyone. Why and how this is so has become a topic of widespread interest. A number of attempts to build human-centered and responsive institutions, more reasonable environments for social intercourse, work, and enjoyment, are now in the hands of those who found it simply impossible to continue the old patterns.

Finally, there is a set of concerns, evident in the aftermath of Vietnam, Watergate, and revelations about the CIA, which aims at restoring the element of responsibility to situations that have tended to exclude responsible conduct. There is a point, after all, where compliance becomes complicity. The twentieth century has made it possible for a person to commit the most ghastly of domestic and foreign crimes by simply living in suburbia and doing a job. The pleas of Lieutenant Calley and Adolf Eichmann—"I just work here"—become the excuse of everyman. Yet for those who perceive the responsibility, when distant deeds are done and the casualties counted, the burdens are gigantic. As Stanley Cavell and Nadezhda Mandelstam have observed, there is a sense in which one comes to feel responsible for literally everything.[26] Evils perpetrated and the good left undone all weigh heavily on one's shoulders. Like Kafka's K. at the door of the castle, the concerned begin a search for someone or something that can be held accountable.

I admit that I have no special name for this collection of projects. *Humanist technology* has been suggested to me, but that seems wide of the mark. At a time in which the industrialization of literature demands catchy paperback titles for things soon forgotten, perhaps it is just as well to leave something truly important unnamed.

The fundamental difference between the two domains, however, can be stated: a difference in insight and commitment. The first, the utilitarian-pluralist approach, sees that technology is problematic in the sense that it now *requires legislation.* An ever-increasing array of rules, regulations, and administrative personnel is needed to maximize the benefits of technological practice while limiting its unwanted maladies. Politics is seen as the process in representative government and interest group interplay whereby such legislation takes shape.

The second approach, disjointed and feeble though it still may be, begins with the crucial awareness that technology in a true sense *is legislation.* It recognizes that technical forms do, to a large extent, shape the basic pattern and content of human activity in our time. Thus, politics becomes (among other things) an active encounter with the specific forms and processes contained in technology.

Along several lines of analysis this book has tried to advance the idea central to all thinking in the second domain—that *technology is itself a political phenomenon.* A crucial turning point comes when one is able to acknowledge that modern technics, much more than politics as conventionally understood, now legislates the conditions of human existence. New technologies are institutional structures within an evolving constitution that gives shape to a new polity, the technopolis in which we do increasingly live. For the most part, this constitution still evolves with little public scrutiny or debate. Shielded by the conviction that technology is neutral and tool-like, a whole new order is built—piecemeal, step by step, with the parts and pieces linked together in novel ways—without the slightest public awareness or opportunity to dispute the character of the changes underway. It is somnambulism (rather than determinism) that characterizes technological politics—on the left, right, and center equally. Silence is its distinctive mode of speech. If the founding fathers had slept through the convention in Philadelphia in 1787 and never uttered a word, their response to constitutional questions before them would have been similar to our own.

Indeed, there is no denying that technological politics as I have described it is, in the main, a set of pathologies. To explain them is to give a diagnosis of how things have gone wrong. But there is no reason why the recognition of technology's intrinsic political aspect should wed us permanently to the ills of the present order. On the contrary, projects now chosen in the second domain bear a common bond with attempts made to redefine an authentic politics and reinvent conditions under which it might be practiced. As a concern for political theory this work has been admirably carried forward by such writers as Hannah Arendt, Sheldon Wolin, and Carole Pateman.[27] In the realm of historical studies it appears as a renewed interest in a variety of attempts—the Paris Communes of 1793 and 1871, nineteenth-century utopian experiments, twentiety-century Spanish anarchism, the founding of worker and community councils in a number of modern revolutions—to create decentralist democratic politics.[28] In contemporary practice it can be seen in the increasingly common efforts to establish worker self-management in factories and bureaucracies, to build self-sufficient communities in both urban and rural settings, and to experiment with modes of direct democracy in places where hierarchy and managerialism had previously ruled.[29]

Taken in this light, it is possible to see technology as legislation and then follow that insight in hopeful directions. An important step comes when one recognizes the validity of a simple yet long overlooked principle: *Different ideas of social and political life entail different technologies for their realization.* One can create systems of production, energy, transportation, information handling, and so forth that are compatible with the growth of autonomous, self-determining individuals in a democratic polity. Or one can build, perhaps unwittingly, technical forms that are incompatible with this end and then wonder how things went strangely wrong. The possibilities for matching political ideas with technological configurations appropriate to them are, it would seem, almost endless. If, for example, some perverse spirit set out deliberately to design a collection of systems to increase the general feeling of powerlessness, enhance the prospects for the dominance of technical elites, create the belief that politics is nothing more than a remote spectacle to be experienced vicariously, and thereby diminish the chance that anyone would take democratic citizenship seriously, what better plan to suggest than that we simply keep the systems we already have? There is, of course, hope that we may decide to do better than that. The challenge of trying to do so now looms as a project open to political science and engineering equally. But the notion that technical forms are merely neutral and that "one size fits all" is a myth that no longer merits the least respect.

Luddism as Epistemology

But what next? Following the normal pattern of twentieth-century writing, I should now rush forward with suggestions and recommendations for how things might be different. What good are analyses, criticisms, and perspectives, some might say, unless they point to positive courses of action?

In view of what we have seen, however, it is not easy simply to take a deep breath and begin spewing forth plans for a better world. The issues are difficult ones. It has not been my aim to make them seem any less difficult than they are. In my experience, virtually all of the remedies proposed are little more than tentative steps in uncertain directions. Goodman's plea for the application of moral categories to technological action, Bookchin's outlines for a liberatory technology, Marcuse's rediscovery of utopian thinking, and Ellul's call to the defiant, self-assertive, free individual—all of these offer us something.[30] But when compared to the magnitude of what is to be overcome, these solutions seem trivial. I could, I suppose, fudge the matter here and seem to be zeroing in on some useful proposals. Having gone this far, the reader can probably predict how it would look.

First, I could say that there is a need to begin the search for new technological forms. Recognizing the often wrong-headed and oppressive character of existing configurations of technology, we should find new kinds of technics that avoid the human problems of the present set. This would mean, presumably, the birth of a new sort of inventiveness and innovation in the physical arrangements of this civilization.

Second, I could suggest that the development of these forms proceed through the direct participation of those concerned with their everyday employment and effects. One major shortcoming in the technologies of the modern period is that those touched by their presence have little or no control over their design or operation. To as great an extent as possible, then, the processes of technological planning, construction, and control ought to be opened to those destined to experience the final products and full range of social consequences.

Third, I might point to the arguments presented here and offer some specific principles to guide further technological construction. One such rule would certainly be the following: *that as a general maxim, technologies be given a scale and structure of the sort that would be immediately intelligible to nonexperts.* This is to say, technological systems ought to be intellectually as well as physically accessible to those they are likely to affect. Another worthy principle would be: *that technologies be built with a high degree of flexibility and mutability.* In other words, we should seek to avoid circumstances in which technological systems impose a permanent, rigid, and irreversible imprint on the lives of the populace. Yet another conceivable rule is this: *that technologies be judged according to the degree of dependency they tend to foster, those creating a greater dependency being held inferior.* This merely recognizes a situation we have seen again and again in this essay. Those who must rely for their very existence upon artificial systems they do not understand or control are not at liberty to change those systems in any way whatsoever. For this reason, any attempt to create new technological circumstances must make certain that it does not discover freedom only to lose it again on the first step.

Finally, I could suggest a supremely important step—that we return to the original understanding of technology as a means that, like all other means available to us, must only be employed with a fully informed sense of *what is appropriate.* Here, the ancients knew, was the meeting point at which ethics, politics, and technics came together. If one lacks a clear and knowledgeable sense of which means are appropriate to the circumstances at hand, one's choice of means can easily lead to excesses and danger. This ability to grasp the appropriateness of means has, I believe, now been pretty thoroughly lost. It has been replaced by an understanding which holds that if a given means can be shown to have a narrow utility, then it ought to be adopted straight off, regardless of its broader implications.[31] For a time, perhaps from the early seventeenth century to the early twentieth, this was a fruitful way of proceeding. But we have now reached a juncture at which such a cavalier disposition will only lead us astray. A sign of the maturity of modern civilization would be its recollection of that lost sense of appropriateness in the judgment of means. We would profit from regaining our powers of selectivity and our ability to say "no" as well as "yes" to a technological prospect. There are now many cases in which we would want to say: "After all a temptation is not very tempting."[32]

I am convinced that measures of this kind point to a new beginning on the problems we have seen.[33] At the same time, these proposals have overtones of utopianism and unreality, which make them less than compelling. It may be that the only innovation I have suggested is to use my hat as a megaphone. There are excellent reasons why *any* call for the taking of a new path or new beginning now falls flat.

Not the least of these is simply the fact that while positive, utopian principles and proposals can be advanced, the real field is already taken. There are, one must admit, technologies already in existence—apparatus occupying space, techniques shaping human consciousness and behavior, organizations giving pattern to the activities of the whole society. To ignore this fact is to take flight from the reality that must be considered. One finds, for example, that in the contemporary discussions those most sanguine about the prospects for tackling the technological dilemma are those who place their confidence in *new* systems to be implemented in the future. Their hope is not that the existing state of affairs will be changed through any direct action, only that certain superior features will be added. In this manner the mass of problems now at hand is skirted.[34]

Another barrier is this: even if one seriously wanted to construct a different kind of technology appropriate to a different kind of life, one would be at a loss to know how to proceed. There is no living body of knowledge, no method of inquiry applicable to our present situation that tells us how to move any differently from the way we already do. Mumford's suggestion that society return to an older tradition of small-scale technics and craftsmanship is not convincing. The world that supported that tradition and gave it meaning has vanished. Where and how techniques of that sort could be a genuine alternative is highly problematic. Certainly a technological revivalism could *add* things to the existing technological stock. But the kind of knowledge that would make a difference is not to be found in decorating the periphery.

In no place is the force of these considerations better exemplified than in the sorry fate of the counterculture of the late 1960s. The belief of those who followed the utopian dream was that by dropping out of the dominant culture and "raising one's consciousness," a better way of living would be produced. In several areas of social fashion—clothing, music, language, drug use—there were some remarkable innovations. But behind the facade of style, a familiar reality still held sway. The basic structures of life, many of them technological structures, remained unchallenged and unchanged. Members of the movement convinced themselves that with a few gestures they had transcended all of that. But all of the networks of practical connections remained intact. The best that was done was to give the existing patterns a hip veneer. Members of the management team began to wear bell-bottoms and medallions.

The lesson, I think, is evident. Even though one commits oneself to ends radically different from those in common currency, there is no real beginning until the question of means is looked straight in the eye. One must take seriously the fact that there are already technologies occupying the available physical and social space and employing the available resources. One must also take seriously the fact that one simply does not yet know how to go ahead to find genuinely new means appropriate to the new "consciousness." No doubt some faced with this realization will simply wish to stop. They will see the virtual necessity of co-optation and the impending disappointment for anyone who tries to resist one's technological fate. Some will find it impossible to do anything else than retreat into despair and blame their plight on "those in power." But if I am not mistaken, the logic of the problem admits at least one more alternative.

In many contemporary writings the response to the idea of autonomous technology reads something like this: "Technology is not a juggernaut; being a human construction it can be torn down, augmented and modified at will."[35] The author of this statement, Dr. Glenn T. Seaborg, would probably be the last person to suggest that any existing technology actually be "torn down." But in his mind, as in many others, the conviction that man still controls technology is rooted in the notion that at any time the whole thing could be taken apart and something better built in its place. This idea, for reasons we have seen all along, is almost pure fantasy. Real technologies do not permit such wholesale tampering. Changes here occur through "invention," "development," "progress," and "growth"—processes in which more and more additions are made to the technological store while some parts are eventually junked as obsolete. The technologies generated are understood to be more or less permanent fixtures. That they might be torn down or seriously tinkered with is unthinkable.

But perhaps Seaborg's idea has some merit. As we have already noted, is not the fundamental business of technics that of taking things apart and putting them together? One conceivable approach to tackling whatever flaws one sees in the various systems of technology might be to begin dismantling those systems. This I would propose not as a solution in itself but as a method of inquiry. The forgotten essence of technical activity, regardless of the specific purpose at hand, might well be revealed by this very basic yet, at the same time, most difficult of steps. Technologies identified as problematic would be taken apart with the expressed aim of studying their interconnections and their relationships to human need. Prominent structures of apparatus, technique, and organization would be, temporarily at least, disconnected and made unworkable in order to provide the opportunity to learn what they are doing for or to mankind. If such knowledge were available, one could then employ it in the invention of radically different configurations of technics, better suited to nonmanipulated, consciously, and prudently articulated ends.

None of this would be necessary if such information were obvious. But at present it is exactly this kind of awareness and understanding that is lacking. Our involvement in advanced technical systems resembles nothing so much as the somnambulist in Caligari's cabinet. Somewhat drastic steps must be taken to raise the important questions at all. The method of carefully and deliberately dismantling technologies, epistemological Luddism if you will, is one way of recovering the buried substance upon which our civilization rests. Once unearthed, that substance could again be scrutinized, criticized, and judged.

I can hear the outcry already. Isn't this man's Luddism simply an invitation to machine smashing? Isn't it mere nihilism with a sharp edge? How can anyone calmly suggest such an awful course of action?

Again, I must explain that I am only proposing a method. The method has nothing to do with Luddism in the traditional sense (the smashing and destroying of apparatus). The much-maligned original Luddites were, of course, merely unemployed workers with a flare for the dramatic. As they scrutinized the mechanization of the textile trade in the industrial revolution, they applied two interesting criteria. Does the new device enhance the quality of the product being manufactured? Does the machine improve the quality of work? If the answer to either question or both is "no," the innovation should not be permitted. Banned from lawful union activity, the Luddites did what they could and unwittingly brought upon themselves a lasting opprobrium.[36]

As best I can tell, there have never been any epistemological Luddites, unless perhaps Paul Goodman was one on occasion. I am not proposing that a sledge hammer be taken to anything. Neither do I advocate any act that would endanger anyone's life or safety. The idea is that in certain instances it may be useful to dismantle or unplug a technological system in order to create the space and opportunity for learning.

The most interesting parts of the technological order in this regard are not those found in the structure of physical apparatus anyway. I have tried to suggest that the technologies of concern are actually *forms of life*—patterns of human consciousness and behavior adapted to a rational, productive design. Luddism seen in this context would seldom refer to dismantling any piece of machinery. It would seek to examine the connections of the human parts of modern social technology. To be more specific, it would try to consider at least the following: (1) the kinds of human dependency and regularized behavior centering upon specific varieties of apparatus, (2) the patterns of social activity that rationalized techniques imprint upon human relationships, and (3) the shapes given everyday life by the large-scale organized networks of technology. Far from any wild smashing, this would be a meticulous process aimed at restoring significance to the question, What are we about?

One step that might be taken, for example, is that groups and individuals would for a time, self-consciously and through advance agreement, extricate themselves from selected techniques and apparatus.

This, we can expect, would create experiences of "withdrawal" much like those that occur when an addict kicks a powerful drug. These experiences must be observed carefully as prime data. The emerging "needs," habits, or discomforts should be noticed and thoroughly analyzed. Upon this basis it should be possible to examine the structure of the human relationships to the device in question. One may then ask whether those relationships should be restored and what, if any, new form those relationships should take. The participants would have a genuine (and altogether rare) opportunity to ponder and make choices about the place of that particular technology in their lives. Very fruitful experiments of this sort could now be conducted with many implements of our semiconscious technological existence, such as the automobile, television, and telephone.

Other possiblities for Luddism as methodology can be found at virtually any point in which social and political institutions depend upon advanced technologies for their effective operation. Persons who, for any reason, wish to alter or reform those institutions—the factory, school, business, public agency—have an alternative open to them that they have previously overlooked. As preparation for changes one may later wish to make, one might try disconnecting crucial links in the organized system for a time and studying the results. There is no getting around the fact that the most likely consequences will be some variety of chaos and confusion. But it is perhaps better to have this out in the open rather than endure the subliminal chaos and confusion upon which many of our most important institutions now rest. Again, these symptoms must be taken as prime data. The effects of systematic disconnection must be taken as an opportunity to inquire, to learn, and seek something better. What is the institution doing in the first place? How does its technological structure relate to the ends one would wish for it? Can one see anything more than to plug the whole back together the way it was before? The Luddite step is necessary if such questions are to be asked in any critical way. It is, perhaps, not too farfetched to suppose that some positive innovations might result from this straightforward challenge to established patterns of institutional life.

By far the most significant of Luddite alternatives, however, requires no direct action at all: the best experiments can be done simply by refusing to repair technological systems as they break down. Many of society's biggest investments at present are those that merely prop up failing technologies. This propping up is usually counted as "growth" and placed in the plus column. We build more and more freeways, larger and larger suburban developments, greater and greater systems of centralized water supply, power, sewers, and police, all in a frantic effort to sustain order and minimal comfort in the sprawling urban complex. Perhaps a better alternative would be to let dying artifice die. One might then begin the serious search, not for something superficially "better" but for totally new forms of sociotechnical existence.

Beyond these few words I have little more to say, for now. This book has taken us on a long path to a conclusion that is actually a beginning. What one does with a beginning is to begin. I realize full well that many of these notions will be counted impractical. But that is precisely the point. I have tried to show that the practical-technical aspect of human activity has been almost totally removed from any concerned and conscious care. Autonomous technology is the part of our being that has been transferred, transformed, and separated from living needs and creative intelligence. Any effort to reclaim this part of human life must at first seem impractical and even absurd.

In this light, the suggestions at the end are not so much a call to action as an attempt to speak to logical problems that arose during the investigation. Given the power of these developments, what might possibly make a difference? My best answer at present is this: if the phenomenon of technological politics is to be overcome, a truly *political technology* must be put in its place. I have tried to give a few outlines of an experimental method that might encourage its birth.

In Mary Shelley's novel, Victor Frankenstein is portrayed as a "modern Prometheus." The young man's inevitable tragedy mirrors an ancient story in which the combined elements of ambition, artifice, pride, and power meet an unfortunate end.

Without doubt the most excellent of Promethean stories, however, is that written by Aeschylus 2,500 years ago. In *Prometheus Bound* we find in luminous, mythical outline many of the themes we have encountered in this essay, for an interesting feature of Aeschylus's treatment of the legend is that it emphasizes the importance of technology in Prometheus's crime against the gods. The fall of man is in Aeschylus's view closely linked to the introduction of science and the arts and crafts. Chained to a desolate rock for eternity, Prometheus describes his plight.

Prometheus I caused mortals to cease foreseeing doom.

Chorus What cure did you provide them with against that sickness?

Prometheus I placed in them blind hopes.

Chorus That was a great gift you gave to men.

Prometheus Besides this, I gave them fire.

Chorus And do creatures of a day now possess bright-faced fire?

Prometheus Yes, and from it they shall learn many crafts.

Chorus These are the charges on which—

Prometheus Zeus tortures me and gives me no respite.[37]

The theft of fire, Aeschylus makes clear, was in its primary consequence the theft of all technical skills and inventions later given to mortals. "I hunted out the secret spring of fire," Prometheus exclaims, "that filled the narthex stem, which when revealed became the teacher of each craft to men, a great resource. This is the sin committed for which I stand accountant, and I pay nailed in my chains under the open sky."[38] As the brash protagonist recounts the specific items he has bestowed upon the human race, it becomes evident that Aeschylus's tale represents the movement of primitive man to civilized society. "They did not know of building houses with bricks to face the sun; know how to work in wood. They lived like swarming ants in holes in the ground, in the sunless caves of the earth."[39] The fire enabled mankind to develop agriculture, mathematics, astronomy, domesticated animals, carriages, and a host of valuable techniques. But Prometheus ends his proud description on a sorry note.

It was I and none other who discovered ships, the sail-driven wagons that the sea buffets. Such were the contrivances that I discovered for men— alas for me! For I myself am *without contrivance to rid myself of my present affliction* [emphasis added].[40]

Prometheus's problem is something like our own. Modern people have filled the world with the most remarkable array of contrivances and innovations. If it now happens that these works cannot be fundamentally reconsidered and reconstructed, humankind faces a woefully permanent bondage to the power of its own inventions. But if it is still thinkable to dismantle, to learn and start again, there is a prospect of liberation. Perhaps means can be found to rid the human world of our self-made afflictions.

Notes

Note to Preface

1. From a speech by Mario Savio, Sit-in Rally, December 2, 1964, Berkeley, California. Transcribed from "Is Freedom Academic?" a documentary of the Free Speech Movement prepared by radio station KPFA.

Notes to Introduction

1. See Martin Brown, ed., *The Social Responsibility of the Scientist* (New York: The Free Press, 1971); John T. Edsall, *Scientific Freedom and Responsibility: A Report of the AAAS Committee on Scientific Freedom and Responsibility* (Washington, D.C.: AAAS, 1975); Joel Primack and Frank von Hippel, *Advice and Dissent: Scientists in the Political Arena* (New York: Basic Books, 1974).

2. See *Organized Social Complexity: Challenge to Politics and Policy*, ed. Todd R. La Porte (Princeton: Princeton University Press, 1975).

3. From the introduction by Allen Ginsberg to Timothy Leary's *Jail Notes* (New York: Douglas Books, 1971).

4. The slogan is usually attributed to Huey P. Newton.

5. Paul Goodman, *The New Reformation: Notes of a Neolithic Conservative* (New York: Random House, 1970), p. 21.

6. Lewis Mumford, *The Myth of the Machine: The Pentagon of Power* (New York: Harcourt Brace Jovanovich, 1970), chap. 13.

7. Ibid., p. 413.

8. Quoted in Hiram Haydn, *The Counter-Renaissance* (New York: Harcourt, Brace, & World, 1950), p. 204.

9. The "values" discussion and its gestures toward relevance seldom arise above the maudlin. See, for example, *The New Technology and Human Values*, ed. John G. Burke (Belmont, California: Wadsworth Publishing Co., 1967).

10. Jacques Ellul, *The Technological Society*, trans. John Wilkinson (New York: Alfred A. Knopf, 1964), p. xxv.

11. Ludwig Wittgenstein, *Philosophical Investigations*, trans. G. E. M. Anscombe (New York: Macmillan, 1958), p. 31e.

Notes to Chapter 1

1. Werner Heisenberg, *Physics and Philosophy* (New York: Harper & Row, 1958), p. 189.

2. John Kenneth Galbraith, *The New Industrial State* (New York: The New American Library, 1968), p. 19.

3. René Dubos, *So Human an Animal* (New York: Scribners, 1968), p. 191. See also Dubos's *Reason Awake* (New York: Columbia University Press, 1970).

4. Dubos, *So Human an Animal*, pp. 231-232.

5. Martin Heidegger, *Discourse on Thinking,* trans. John M. Anderson and E. Hans Freund (New York: Harper & Row, 1966), p. 51.

6. Charles Reich, *The Greening of America* (New York: Random House, 1970), p. 134.

7. Alvin Toffler, *Future Shock* (New York: Random House, 1970), p. 394.

8. Albert Speer, *Inside the Third Reich,* trans. Richard and Clara Winston (New York: Macmillan, 1970), p. 521.

9. Ibid.

10. Geoffrey Barraclough, "Hitler's Master Builder," *New York Review of Books* 15, no. 12, January 7, 1971, pp. 6-15. "What distinguished the Nazis was their primitivism," Barraclough comments, "not their modernity, and the tendency to sheer away from this unpleasant fact and blame everything instead on technology and the depersonalization of man is a typical piece of German double-think." (ibid., p. 14).

11. Jacques Ellul, *The Technological Society,* trans. John Wilkinson (New York: Alfred A. Knopf, 1964), p. 14.

12. Bruno Bettelheim, *The Informed Heart: Autonomy in a Mass Age* (New York: The Free Press, 1960).

13. Immanuel Kant, *Critique of Practical Reason,* trans. Lewis White Beck (Indianapolis and New York: Bobbs-Merrill, 1956), pp. 33-34.

14. Ellul, *Technological Society,* p. 138.

15. B. F. Skinner, *Beyond Freedom and Dignity* (New York: Alfred A. Knopf, 1971), p. 205.

16. Ibid., p. 3.

17. Ibid., p. 24.

18. George Kateb, *Utopia and Its Enemies* (New York: The Free Press, 1963), p. 109.

19. Ibid., p. 108.

20. Seymour Melman, *Pentagon Capitalism: The Political Economy of War* (New York: McGraw-Hill, 1970), p. 12. Professor Melman's comments are directed at the views of nuclear physicist Ralph E. Lapp in *The Weapons Culture* (Baltimore: Penguin Books, 1968), p. 19: "The United States has institutionalized its arms-making to a point where there is grave doubt that it can control this far-flung apparatus."

21. See Ernst Nolte, *Three Faces of Fascism,* trans. Leila Vennewitz (New York: New American Library, 1969), pp. 509-510, parts 4, 5; See also Fritz Stern, *The Politics of Cultural Despair* (New York: Doubleday & Company, Anchor Books, 1965), introduction and pp. 1-22, 52-60, 151-177. Martin Heidegger was attracted to nazism for a brief period, apparently under the belief that the movement

represented an "encounter between global technology and modern man." See Hannah Arendt's "Martin Heidegger at Eighty," *New York Review of Books* 18, no. 6, October 21, 1971, pp. 50-54.

22. E. J. Dijksterhuis, *The Mechanization of the World Picture,* trans. C. Dikshoorn (Oxford: Oxford University Press, 1961), p. 74.

23. Aristotle, *Politics,* ed. and trans. Ernest Barker (New York: Oxford University Press, 1958), p. 10.

24. Francis Bacon, *Novum Organum,* in *Selected Writings,* ed. Hugh G. Dick (New York: The Modern Library, 1955), p. 499.

25. Ibid., p. 537.

26. Ibid., pp. 537-539.

27. Ibid.

28. Ibid.

29. Ibid.

30. Geoffrey Chaucer, *The Canterbury Tales,* trans. Nevill Coghill (Baltimore: Penguin Books, 1960), pp. 472-473, 477.

31. A. Rupert Hall, *From Galileo to Newton: 1630-1720* (New York: Harper & Row, 1963), p. 329.

32. T. K. Derry and Trevor I. Williams, *A Short History of Technology* (London: Oxford University Press, 1970), pp. 702-703.

33. Aristotle, *Metaphysics,* trans. Richard Hope (Ann Arbor: The University of Michigan Press, 1960), pp. 3-4.

34. Hannah Arendt, *The Human Condition* (New York: Doubleday & Company, Anchor Books, 1959), pp. 197-206.

35. H. L. Nieburg, *In the Name of Science* (Chicago: Quadrangle, 1966), p. v.

36. See Richard R. Landers, *Man's Place in the Dybosphere* (Englewood Cliffs: Prentice-Hall, 1966), p. 207, "Technology in itself is neutral and should not be labeled 'good' or 'bad.' It is the uses to which we put new scientific developments that enhance or degrade personal well being and prosperity." See also R. J. Forbes, *The Conquest of Nature: Technology and Its Consequences* (New York: New American Library, 1968), pp. 107-114.

37. Philip E. Slater, *The Pursuit of Loneliness* (Boston: Beacon Press, 1970), pp. 12-13.

38. See Herbert L. Sussman, *Victorians and the Machine: The Literary Response to Technology* (Cambridge, Mass.: Harvard University Press, 1968).

39. Nathaniel Hawthorne, "The Artist of the Beautiful," in *Selected Tales and Sketches* (New York: Holt, Rinehart and Winston, 1963), p. 259.

40. Ibid., p. 267.

41. Edgar Allan Poe, "Maelzel's Chess Player," in *The Complete Tales and Poems of Edgar Allan Poe* (New York: The Modern Library, 1938), p. 433.

42. Kurt Vonnegut, Jr., *Player Piano* (New York: Avon Books, 1967), p. 18.

43. Ibid., pp. 37-38.

44. Hawthorne, *Selected Tales*, p. 263.

45. Ibid., p. 264.

46. E. M. Forster, "The Machine Stops," in *Of Men and Machines*, ed. Arthur O. Lewis, Jr. (New York: E. P. Dutton, 1963), p. 266.

47. Ibid., p. 279.

48. Ibid., p. 263.

49. Ibid., p. 265.

50. Karl Marx, *Capital*, Vol. I, 3d ed., trans. Samuel Moore and Edward Aveling (New York: The Modern Library, 1906), p. 416.

51. Ibid., pp. 461-462.

52. Ibid., p. 416.

53. Karl Marx, *Economic and Philosophical Manuscripts*, in *Karl Marx: Early Writings*, trans. and ed. T. B. Bottomore (New York: McGraw-Hill, 1964), pp. 127-128.

54. Ibid., p. 127.

55. Karl Marx and Friedrich Engels, *The German Ideology*, parts I & II, ed. R. Pascal (New York: International Publishers, 1964), pp. 127-128.

56. Marx, *Manuscripts*, p. 122.

57. Ibid.

58. Marx, *Capital*, p. 462.

59. Marx and Engels, *German Ideology*, p. 22-23.

60. Ibid., p. 70.

61. Ibid., p. 67-68.

62. Karl Marx, *Selected Writings in Sociology and Social Philosophy*, trans. T. B. Bottomore and ed. T. B. Bottomore and Maximilien Rubel (New York: McGraw-Hill, 1964), p. 108.

63. See Karl Marx, *Grundrisse: Foundations of the Critique of Political Economy*, trans. Martin Nicolaus (Harmondsworth: Penguin Books, 1973), pp. 196-197. Here Marx makes much the same argument with regard to money and circulation as he does elsewhere in considering the dialectic of men and machines: "The social relation of individuals to one another as a power over the individuals which has become autonomous, whether conceived as a natural force,

as chance or in whatever other form, is a necessary result of the fact that the point of departure is not the free social individual."

64. Marx, *Capital,* pp. 462-463.

65. Ibid., p. 462.

66. Marx, *Manuscripts,* p. 130.

67. Jacques Ellul, *La Technique ou l'enjeu du siècle* (Paris: Librairie Armand Colin, 1954).

Notes to Chapter 2

1. Henry Adams, *The Education of Henry Adams* (New York: The Modern Library, 1931), p. 380.

2. Ibid.

3. Quoted in Leo Marx's *The Machine in the Garden: Technology and the Pastoral Ideal in America* (New York: Oxford University Press, 1964), p. 350.

4. For an excellent analysis of the computer as a metaphor of this kind, see Joseph Weizenbaum, *Computer Power and Human Reason: From Judgment to Calculation* (San Francisco: W. H. Freeman and Company, 1976).

5. Reinhard Bendix attempts to distinguish between the major terms used to categorize this historical movement in his *Nation-Building and Citizenship: Studies of our Changing Social Order* (Garden City: Doubleday & Company, Anchor Books, 1969), pp. 1-18.

6. Walt Whitman Rostow, *Politics and the Stages of Growth* (New York, Cambridge University Press, 1971), p. 56.

7. Ibid., p. 3.

8. Wilbert E. Moore, *Social Change* (Englewood Cliffs: Prentice-Hall, 1963), p. 89.

9. Jacques Ellul, *The Technological Society,* trans. John Wilkinson (New York: Alfred A. Knopf, 1964), p. 134.

10. Adams, *Education,* p. 490.

11. Ibid., p. 493.

12. Roderick Seidenberg, *Post-Historic Man* (Boston: Beacon Press, 1957), pp. 234-235. See also Norbert Wiener, *The Human Use of Human Beings: Cybernetics and Society* (New York: Avon Books, 1967), pp. 20-21.

13. David S. Landes, *The Unbound Prometheus: Technological Change and Industrial Development in Western Europe from 1750 to the Present* (Cambridge: Cambridge University Press, 1969), p. 3.

14. C. E. Black, *The Dynamics of Modernization: A Study in Comparative History* (New York: Harper & Row, 1967), p. 7.

15. Clark Kerr, *Industrialism and Industrial Man* (New York: Oxford University Press, Galaxy Books, 1964), p. 15.

16. Ibid.

17. Ibid., p. 17.

18. David E. Apter, *The Politics of Modernization* (Chicago: University of Chicago Press, 1965), p. 1.

19. Myron Weiner, ed., *Modernization: The Dynamics of Growth* (New York: Basic Books, 1966), p. 2.

20. Ibid., p. v.

21. For an assessment of the attitudes of different social classes toward the industrial revolution see Landes, *Unbound Prometheus,* pp. 121-123.

22. Victor C. Ferkiss, *Technological Man: The Myth and the Reality* (New York: George Braziller, 1969), p. 87.

23. There are any number of books that show the aspect of technical and scientific change influenced by human personality and individual idiosyncrasy. See, for example, William Rodgers, *Think: A Biography of the Watsons and IBM* (New York: New American Library, 1969), and James D. Watson, *The Double Helix* (New York: New American Library, 1969), especially the touching passage on p. 104 in which Watson and Francis Crick drink a toast to the failure of their rival, Linus Pauling.

24. Kerr, *Industrialism,* pp. 221-223.

25. Apter, *Politics,* pp. 3, 10.

26. Rostow, *Politics,* p. 176.

27. Ibid.

28. Ibid., p. 177.

29. Ellul, *Technological Society.*

30. Karl Marx, *Capital,* Vol. I, 3d ed., trans. Samuel Moore and Edward Aveling (New York: The Modern Library, 1906), p. 406.

31. Ellul, *Technological Society,* p. 134.

32. Ibid., p. 85.

33. Ibid., p. 78.

34. See Richard R. Landers, *Man's Place in the Dybosphere* (Englewood Cliffs: Prentice-Hall, 1966), p. 243, for a wonderful example of this kind of thinking: "Objectively," Landers confesses, "I cannot help feeling that, relatively, man is declining and machines are growing. Not in the sense that machines are 'taking over' in a robot uprising, but in the sense that while man may only be around for one million years more, machines may be around for two million years."

35. Arthur C. Clarke, *Profiles of the Future* (New York: Bantam Books, 1964), pp. 212-227.

36. Ellul, *Technological Society,* p. 85.

37. Ibid., pp. 7-10.

38. Ibid., p. 86.

39. Ibid.

40. Ibid.

41. Ibid.

42. Ibid., p. 87.

43. Ibid.

44. Ibid., p. 91.

45. Ibid., p. 87.

46. Ibid., p. 90.

47. Ibid.

48. Ibid.

49. Ibid., p. 93.

50. Emile Durkheim, *The Rules of Sociological Method,* 8th ed., trans. Sarah A. Solovay and John H. Mueller and ed. George E. G. Catlin (New York: The Free Press, 1964), p. 103.

51. Ellul, *Technological Society,* p. xxviii.

52. Ibid.

53. Alvin W. Gouldner, *The Coming Crisis of Western Sociology* (New York: Basic Books, 1970), p. 52.

54. Ellul, *Technological Society,* pp. 92-93.

55. For a good discussion of the viewpoints of Karl Popper and Thomas Kuhn on the history of science, see *Criticism and the Growth of Knowledge,* edited by Imre Lakatos and Alan Musgrave (Cambridge: Cambridge University Press, 1970). See also, Thomas Kuhn's *The Structure of Scientific Revolutions,* 2d ed. (Chicago: University of Chicago Press, 1970).

56. See Donald A. Schon, *Technology and Change* (New York: Delacorte Press, 1967).

57. Donald A. Schon, *Beyond the Stable State* (New York: Random House, 1971), p. 24.

58. George Kubler, *The Shape of Time: Remarks on the History of Things* (New Haven: Yale University Press, 1962), p. 7.

59. A. L. Kroeber, *Anthropology: Culture Patterns and Processes* (New York: Harbinger Press, 1963), p. 150.

60. Ibid., pp. 172-173.

61. Ibid.

62. Durkheim, *Rules,* p. 28.

63. Ibid., p. 110.

64. Werner Heisenberg, *Physics and Beyond,* trans. Arnold J. Pomerans (New York: Harper & Row, 1971), p. 193.

65. Ibid.

66. Ibid., p. 194.

67. Ibid.

68. Ibid., p. 195.

69. Ibid.

70. By its very nature, of course, evidence of actual suppression of scientific or technical work for reasons of conscience might never become known. I have encountered scientists who claimed to have stopped lines of research that they feared might have pernicious military applications. And perhaps there exists something like an anonymous history of technical paths that, through prudence, were not taken.

71. Heisenberg, *Physics,* p. 195.

72. Norbert Wiener, *Cybernetics* (Cambridge, Mass.: The MIT Press, 1965), pp. 28-29.

73. Kubler, *Shape of Time,* p. 36.

74. From *In the Matter of J. Robert Oppenheimer: Transcript of Hearing before Personnel Security Board and Texts of Principal Documents and Letters* (Cambridge, Mass.: The MIT Press, 1971), p. 251.

75. Benjamin Farrington, *Greek Science* (Baltimore: Penguin Books, 1961), p. 303.

76. See, for example, Herman Kahn and Anthony J. Wiener, *The Year 2000: A Framework for Speculation* (New York: Macmillan, 1967); Dennis Gabor, *Inventing the Future* (New York: Alfred A. Knopf, 1964); Daniel Bell, "The Year 2000: The Trajectory of an Idea," *Daedalus* 96 (1967), pp. 639-651.

77. Robert Theobald, "Cybernetics and the Problems of Social Reorganization," in *The Social Impact of Cybernetics,* ed. Charles R. Dechert (New York: Simon & Schuster, 1966), p. 39.

78. See the definitions given in *The American Heritage Dictionary of the English Language* (Boston: Houghton Mifflin, 1970).

79. Leslie White, *The Science of Culture* (New York: Farrar, Straus & Giroux, 1949), p. 366.

80. Emmanuel G. Mesthene, *Technological Change: Its Impact on Man and Society* (New York: New American Library, 1970), p. 20.

81. Bendix in *Nation-Building* criticizes Thorstein Veblen's "technological determinism" and proceeds to develop what he holds to be a more balanced view of social change.

82. Lynn White, *Medieval Technology and Social Change* (New York: Oxford University Press, 1966), p. 28.

83. See William F. Ogburn, *Social Change* (New York: Huebsch, 1922), and William F. Ogburn and M. F. Nimkoff, *Technology and the Changing Family* (Boston: Houghton Mifflin, 1955).

84. Karl Marx, *Selected Writings in Sociology and Social Philosophy,* trans. T. B. Bottomore and ed. T. B. Bottomore and Maximilien Rubel (New York: McGraw-Hill, 1964), p. 64.

85. Karl Marx and Friedrich Engels, *The German Ideology,* parts 1 and 2, ed. R. Pascal (New York: International Publishers, 1964), p. 7.

86. Marx, *Selected Writings,* p. 57.

87. The richness of Marx's view of the range of influences that affect forms of social life is particularly evident in his discussion of precapitalist societies. Here he acknowledges the role of kinship, religion, and other factors while, of course, emphasizing relations of property. See his *Grundrisse: Foundations of the Critique of Political Economy* (Harmondsworth: Penguin Books, 1973), pp. 471-514.

88. Marx and Engels, *German Ideology,* p. 18.

89. Karl Marx, *Poverty of Philosophy* (New York: International Publishers, 1963), p. 109.

90. Ibid., p. 63.

91. Ibid., p. 122.

92. Marx and Engels, *German Ideology,* p. 8.

93. Marx, *Poverty of Philosophy,* p. 139.

94. Georg Lukács, *History and Class Consciousness: Studies in Marxist Dialectics,* trans. Rodney Livingstone (Cambridge, Mass.: The MIT Press, 1971).

95. Marx, *Poverty of Philosophy,* pp. 181-182.

96. Marx and Engels, *German Ideology,* p. 22.

97. See Friedrich Engels, *Anti-Dühring,* trans. Emile Burns and ed. C. P. Dutt (New York: International Publishers, 1939), and Friedrich Engels, *Dialectics of Nature,* trans. and ed. Clemens Dutt (New York: International Publishers, 1940).

98. Marx, *Selected Writings,* p. 63.

99. Marx, *Poverty of Philosophy,* p. 181.

100. Ibid., p. 141.

101. Karl Marx, *Grundrisse,* ed. and trans. David McLellan (New York: Harper & Row, 1971), p. 125.

102. Ibid., pp. 94-95.

103. See Marx and Engels, *German Ideology,* p. 17. Shlomo Avineri's study of Marx's thought points to the idea of an infinity of human needs expressed in Marx's argument. *The Social and Political Thought of Karl Marx* (Cambridge: Cambridge University Press, 1970), pp. 80-82.

104. Marx, *Capital,* p. 399.

105. Pertti J. Pelto, *The Snowmobile Revolution: Technology and Social Change in the Arctic* (Menlo Park, California: Cummings Publishing Co., 1973).

106. Ibid., pp. 178-179.

107. Ferkiss, *Technological Man,* p. 272.

108. Ibid., p. 246.

109. Robert L. Heilbroner, *Between Capitalism and Socialism* (New York: Vintage, 1970), p. 163, emphasis deleted.

110. Of course there are "winners" as well, but as shall be evident, their success is actually assumed and need not be anticipated or foreseen through any extraordinary means.

111. *Technology: Processes of Assessment and Choice:* National Academy of Sciences, Committee on Science and Astronautics, U.S. House of Representatives (Washington, D.C.: Government Printing Office, 1969), pp. 74-77. See also *A Study of Technology Assessment,* Report of the Committee on Public Engineering Policy, National Academy of Engineering, Committee on Science and Astronautics, U.S. House of Representatives (Washington, D.C.: Government Printing Office, 1969). My views of the assessment business are found in "On Criticizing Technology," *Public Policy* 20 (Winter 1972): 35-59.

113. From *Greek Lyrics,* 2d. ed., trans. Richmond Lattimore (Chicago: University of Chicago Press, 1960), p. 20.

114. Niccolò Machiavelli, *The Prince,* trans. Luigi Ricci (New York: New Ameri-American Library, 1952), p. 120.

115. Marcus Aurelius, *Meditations* (Chicago: Henry Regnery, 1956), p. 79.

116. Jean-Paul Sartre, *Search for a Method,* trans. Hazel E. Barnes (New York: Alfred A. Knopf, 1963), p. 47.

117. Hannah Arendt, *The Human Condition* (Chicago: University of Chicago Press, 1958), p. 190.

118. Ibid.

119. Ibid., pp. 323-324.

120. Ibid., pp. 231-232.

121. Friedrich Nietzsche, *The Will to Power,* trans. Walter Kaufmann and R. J. Hollingdale and ed. Walter Kaufmann (New York: Random House, 1967), p. 164.

122. Arendt, *Human Condition,* p. 233.

123. See, for example, Robert Alex Baron's *The Tyranny of Noise* (New York: Harper & Row, 1971).

124. National Academy of Sciences, *Technology,* p. 15.

125. Ibid., p. 11.

126. Ayn Rand, *The New Left: The Anti-Industrial Revolution* (New York: New American Library, 1971), p. 131.

127. James D. Thompson, *Organizations in Action* (New York: McGraw-Hill, 1967), p. 14.

128. Ibid., p. 40.

129. Ellul, *Technological Society,* pp. 94-133.

130. Landes, *Unbound Prometheus,* p. 3.

131. Ibid., p. 546.

132. Ibid., p. 2.

133. Rostow, *Politics and the Stages of Growth,* p. 58.

134. Ibid., pp. 58-59.

135. Landes, *Unbound Prometheus,* chap. 3.

136. John Kenneth Galbraith, *The New Industrial State* (New York: New American Library, 1968), chap. 2.

137. Ellul, *Technological Society,* p. 125.

Notes to Chapter 3

1. See, for example, Edward O. Wilson, *Sociobiology: The New Synthesis* (Cambridge, Mass: Harvard University Press, 1975).

2. Lewis Mumford, *The Myth of the Machine: The Pentagon of Power* (New York: Harcourt Brace Jovanovich 1970), p. 155.

3. Lewis Mumford, *The Transformations of Man* (New York: Harper & Row, 1956).

4. Lewis Mumford, *The Myth of the Machine: Technics and Human Development* (New York: Harcourt, Brace and World, 1967).

5. Barnett Newman, "The First Man Was an Artist," *The Tiger's Eye* 1 (October, 1947), p. 59.

6. Ibid.

7. Ibid.

8. See Sherwood L. Washburn, "Tools and Human Evolution," in *Scientific Technology and Social Change,* ed. Gene I. Rochlin (San Francisco: W. H. Freeman and Company, 1974), pp. 11-23.

9. Newman, "First Man," p. 59.

10. Hannah Arendt, *The Human Condition* (Chicago: University of Chicago Press, 1958), pp. 9-10.

11. Lynn White, Jr., "The Historical Roots of Our Ecologic Crisis," in *Philosophy and Technology: Readings in the Philosophical Problems of Technology,* ed. Carl Mitcham and Robert Mackey (New York: The Free Press, 1972), p. 260.

12. Max Weber, *The Protestant Ethic and the Spirit of Capitalism,* trans. Talcott Parsons (New York: Charles Scribner's Sons, 1930). For another view of this issue see Robert K. Merton, *Science, Technology and Society in Seventeenth-Century England* (New York: Harper & Row, 1970). Merton's bibliography lists works published on this topic up to 1970. See also *Bibliography of the Philosophy of Technology,* comp. Carl Mitcham and Robert Mackey (Chicago: University of Chicago Press, 1973), pp. 123-141.

13. White, "Historical Roots," p. 264.

14. Ibid., p. 261.

15. Ibid., pp. 262-263.

16. Ibid., pp. 262-263.

17. Genesis 1:28.

18. Matthew 5:5.

19. See David S. Landes, *The Unbound Prometheus: Technological Change and Industrial Development in Western Europe from 1750 to the Present* (Cambridge: Cambridge University Press, 1969), pp. 21-25. Landes weighs a range of cultural sources for European attitudes toward science and industry. He concludes that "the urge to mastery grew with time and fed on success, for every achievement was justification for the pretention; while the moral force of the Church's opposition waned with its temporal power and its own insecurity in the face of a triumphant materialism."

20. John Passmore, *Man's Responsibility for Nature: Ecological Problems and Western Traditions* (London: Duckworth, 1974), ix.

21. Genesis 1:25.

22. Passmore, *Man's Responsibility,* p. 12.

23. Ibid., p. 20.

24. Max Horkheimer, *Eclipse of Reason* (New York: Seabury Press, 1974), p. 21.

25. Ibid., p. 176. For an analysis of the role that the domination-of-nature thesis played in the development of the thinking of the Frankfurt school, see Martin Jay, *The Dialectical Imagination: A History of the Frankfurt School and the Institute of Social Research, 1923-1950* (Boston: Little, Brown, 1973), chap. 8.

26. Max Horkheimer and Theodor W. Adorno, *Dialectic of Enlightenment,* trans. John Cumming (London: Allen Lane, 1973), pp. 43-80.

27. In the context of the present inquiry I shall not be able to consider the range of positions taken by Adorno, Horkheimer, and others of the Frankfurt school on questions of nature, technology, and domination. However, insofar as their writings involve a search for ultimate origins of a pathological condition in Western civilization's dealing with nature, their work is subject to many of the criticisms I have raised with regard to White and Passmore.

28. Jacques Ellul, *The Technological Society,* trans. John Wilkinson (New York: Alfred A. Knopf, 1964), p. 146.

29. Ibid., p. xxix.

30. Ibid.

31. Ibid., p. 66.

32. Ibid., p. 29.

33. Ibid., p. 29.

34. Ibid.

35. Ibid.

36. Ibid., p. 34.

37. Ibid., p. 35.

38. Ibid., p. 37.

39. Ibid.

40. Lewis Mumford, *Technics and Civilization* (New York: Harcourt, Brace & World, 1963), pp. 242-249 and Chaps. I, III and IV.

41. Plato, *Gorgias,* trans. W. C. Helmbold (New York: Bobbs-Merrill Co., 1952), p. 97.

42. See Friedrich Klemm, *A History of Western Technology,* trans. Dorothea Waley Singer (Cambridge, Mass: The MIT Press, 1964), pp. 153-159.

43. Ellul, *Technological Society,* pp. 41-42.

44. Ibid., p. 41.

45. Ibid., p. 48.

46. See Ellul's *Presence of the Kingdom* (New York: Seabury Press, 1967). In a chapter entitled "The Christian in the World," he says (p. 16), "We have no right to accustom ourselves to this world, nor to hide it from ourselves with Christian

illusions. Living in the world we are living in the domain of the Prince of this world, of Satan, and all around us we constantly see the action of this Prince, and the result of the state of sin in which we are all placed without exception, because in spite of all our efforts and our piety we share in the sin of the world." As one compares Ellul's sociological and theological writings, it becomes clear that, in his eyes, sin and *la technique* are virtual equivalents. Technology is Satan's modern variety of temptation, his way of claiming the soul of thoroughly secularized men. What, after all, could be a greater temptation than the promise of total effectiveness and efficiency in all one's worldly affairs?

47. Ellul, *Technological Society,* p. 51-52.

48. Ibid., p. 43.

49. Ibid.

50. Ellul says the following about his early intellectual development: "I was not brought up in an especially Christian family, and had only a very remote knowledge of Christianity in my childhood. On the other hand, my family was rather poor and I spent all my youth in the midst of the people of the docks at Bordeaux. I began to earn my own living when I was sixteen and continued to do so while completing my university studies. When I was nineteen, I read, by chance, Marx's *Capital.* I was enthusiastic about it. It answered all the questions that I had been asking myself. I became 'Marxist' and devoted a great deal of my time to a study of his writings. But I was disappointed with the Communists, who seemed to me to be very far from Marx, and I never entered the Party. Around twenty-two years of age, I was also reading the Bible, and it happened that I was converted with a certain 'brutality'!

"From that time on, the great problem for me was to know if I could be Marxist and Christian. On the philosophical plane, I realized very quickly that I could not, and so chose decisively for faith in Jesus Christ. But what Marx had brought to me was a certain way of 'seeing' the political, economic and social problems—a method of interpretation, a sociology. So it did not seem impossible to utilize this, starting with the Christian faith. I could not accept the view that there should be a Christian faith without social and political consequences. On the other hand, however, I saw clearly that one could not deduce directly from the Biblical texts political or social consequences valid for our epoch. It seemed to me that the method of Karl Marx (but not of the Communists!) was superior to all that I had encountered elsewhere." (James Y. Holloway, ed., *Introducing Jacques Ellul* [Grand Rapids, Mich.: Eerdmans Publishing Company, 1970], p. 5).

51. Ellul, *Technological Society,* pp. 54-55, 222.

52. Louis Hartz, *The Liberal Tradition in America* (New York: Harcourt, Brace & World, 1955).

53. Ibid., p. 128.

54. The text of Lord Byron's speech to the House of Lords can be found in Arthur O. Lewis, Jr., ed., *Of Men and Machines* (New York: E.P. Dutton, 1963), pp. 198-204.

55. Ellul, *Technological Society,* p. 19.

56. Wylie Sypher, *Literature and Technology: The Alien Vision* (New York: Random House, Vintage Books, 1971), p. xvi.

57. See Edmund Husserl, *Phenomenology and the Crisis of Philosophy,* trans. Quentin Lauer (New York: Harper Torchbooks, 1965); Martin Heidegger, *What Is Called Thinking?,* trans. Fred D. Wieck and J. Glenn Gray (New York: Harper & Row, 1968), and "Die Frage nach der Technik," in *Vorträge und Aufsätze* (Pfullingen: Neske, 1954); Max Scheler, *Die Wissenformen und die Gesellschaft,* in *Gesammelte Werke,* Band 8, 2d ed. (Bern: Francke Verlag, 1960).

58. Ellul, *Technological Society,* p. 428.

59. Heidegger, *What is Called Thinking?,* p. 235.

60. White, "Historical Roots," pp. 264. See also Passmore's recommendations in chap. 7, "Removing the Rubbish," of *Man's Responsibility.*

61. William Leiss, *The Domination of Nature* (Boston: Beacon Press, 1974), p. 193.

Notes to Chapter 4

1. Francis Bacon, *New Atlantis,* in *Selected Writings of Francis Bacon*, ed. Hugh G. Dick (New York: The Modern Library, 1955), p. 554.

2. Ibid., P. 574.

3. Ibid., P. 561.

4. Ibid., p. 583.

5. Ibid., p. 567.

6. Ibid., p. 549.

7. Ibid., p. 584.

8. Ibid., p. 573.

9. Ibid., p. 572.

10. Lewis Mumford argues this point in his discussion of Bacon in *The Myth of the Machine: The Pentagon of Power* (New York: Harcourt Brace Jovanovich, 1970), p. 106. Mumford, p. 166, traces the birth of "the Power Complex: a new constellation of forces, interests, and motives which eventually resurrected the ancient megamachine, and gave it a more perfect technological structure, capable of planetary and even interplanetary extension," back to the Baconian model. He gives much less emphasis to a person equally important in the founding of the successful world view of modern science and technology, Isaac Newton. For an

interesting discussion of Newton's role in the politics of knowledge of the Royal Society, see Frank E. Manuel, "Newton as Autocrat of Science," *Daedalus,* vol. 97, no. 3 (Summer 1968): 969-1001.

11. For example, Spencer Klaw, *The New Brahmins: Scientific Life in America* (New York: Apollo Books, 1969).

12. For a fruitful contrast to technocratic notions of power and authority, see Hannah Arendt's *On Violence* (New York: Harcourt, Brace & World, 1970), pp. 35-56, and *Between Past and Future* (New York: The Viking Press, 1968), chap. 3.

13. Henri de Saint-Simon, *Social Organization, the Science of Man and Other Writings,* ed. and trans. Felix Markham (New York: Harper & Row, 1964), p. 78.

14. For a summary of such plans in the context of Saint-Simon's life and philosophy, see Frank E. Manuel's *The New World of Henri Saint-Simon* (Notre Dame, Indiana: University of Notre Dame Press, 1963), chaps. 24-27.

15. Saint-Simon, *Social Organization,* p. 6.

16. An excellent review of Wells's science fiction is given in Herbert L. Sussman's *Victorians and the Machine: The Literary Response to Technology* (Cambridge, Mass.: Harvard University Press, 1968), chap. 6.

17. H. G. Wells, *When the Sleeper Wakes,* in *Three Prophetic Novels* (New York: Dover Publications, 1960).

18. H. G. Wells, *A Modern Utopia* (Lincoln: University of Nebraska Press, 1967).

19. F. W. Taylor, *The Principles of Scientific Management* (New York: Harper & Row, 1947).

20. See Robert Boguslaw, *The New Utopians: A Study of System Design and Social Change* (Englewood Cliffs: Prentice-Hall, 1965), and Daniel Bell, *The Coming of Post-Industrial Society: A Venture in Social Forecasting* (New York: Basic Books, 1973).

21. Saint-Simon, *Social Organization,* p. 60.

22. Thorstein Veblen, *The Engineers and the Price System* (New York: The Viking Press, 1954), p. 40.

23. Ibid.

24. Ibid., p. 39.

25. Ibid., p. 54.

26. Ibid., p. 52.

27. Thorstein Veblen, *The Portable Veblen,* ed. Max Lerner (New York: The Viking Press, 1948), p. 443.

28. Veblen, *The Engineers,* p. 57.

29. Oswald Spengler, *The Decline of the West,* trans. Charles Francis Atkinson (New York: Alfred A. Knopf, 1928), 2:504.

30. Ibid.

31. Ibid., pp. 504-505.

32. Oswald Spengler, *Man and Technics,* trans. Charles Francis Atkinson (New York: Alfred A. Knopf, 1932), p. 63. Spengler's ideas on technocracy are mixed with an unashamed racism. He predicted a battle of the "coloured" peoples (including the Russians) against the white peoples of the world. "Even on the present scale our technical processes and installations, if they are to be maintained, require, let us say a hundred thousand outstanding brains, as organizers and discoverers and engineers. These must be strong—nay, even creative—talents, enthusiasts for their work, and formed for it by a steeling of years' duration at great expense. Actually, it is just this calling that has for the last fifty years irresistibly attracted the strongest and ablest of white youth" (ibid., p. 96).

33. Ibid., pp. 96-97, 102-103.

34. See Vilfredo Pareto, *The Mind and Society* (London: Jonathan Cape, 1935); Gaetano Mosca, *The Ruling Class* (New York: McGraw-Hill, 1939); Robert Michels, *Political Parties* (Glencoe: The Free Press, 1949); C. Wright Mills, *The Power Elite* (New York: Oxford University Press, 1956); Floyd Hunter, *Community Power Structure* (Chapel Hill: University of North Carolina Press, 1953); G. William Domhoff, *Who Rules America?* (Englewood Cliffs: Prentice-Hall, 1967).

35. Quoted by T. B. Bottomore, *Elites and Society* (Baltimore: Penguin Books, 1966), p. 8.

36. Quoted by Seymour Melman in *Pentagon Capitalism* (New York: McGraw-Hill, 1970), p. 238. See also appendix A, pp. 231-234, for an Eisenhower memorandum of 1946 that expresses a much different sentiment. "Scientists and industrialists are more likely to make new and unsuspected contributions to the development of the Army if detailed directions are held to a minimum."

37. Jean Meynaud looks for technocracy among the ranks of the "technologists." "The most commonly accepted view of a technologist is of a specialist who, by training or experience, has a thorough knowledge of a particular field or subject." Jean Meynaud, *Technocracy,* trans. Paul Barnes (New York: The Free Press, 1964). "Technologists are," Meynaud observes, "potential technocrats—the extent of their ultimate powers of intervention depends especially on their professional status" (ibid., p. 29). Meynaud distinguished between technocrats and bureaucrats and between various possible kinds of technocracy. The conceptual categories he develops, however, are neither incorrect nor particularly useful.

38. There is a large and growing literature on the role of scientists and technically trained experts in American government. See, for example, A. Hunter Dupree,

Science in the Federal Government (Cambridge, Mass.: Belknap Press, 1957); Daniel S. Greenberg, *The Politics of Pure Science* (New York: New American Library, 1968); Eugene B. Skolnikoff, *Science, Technology and American Foreign Policy* (Cambridge, Mass.: The MIT Press, 1967).

39. Don K. Price, *The Scientific Estate* (Cambridge, Mass.: Harvard University Press, 1965).

40. Victor C. Ferkiss, *Technological Man: The Myth and the Reality* (New York: George Braziller, 1969), p. 175.

41. Meynaud, *Technocracy,* p. 296.

42. Henry Elsner, Jr., *The Technocrats: Prophets of Automation* (Syracuse: Syracuse University Press, 1967).

43. Ludwig Wittgenstein, *Philosophical Investigations,* trans. G. E. M. Anscombe (New York: Macmillan, 1953), p. 48e, section no. 115. "A *picture* held us captive. And we could not get outside it, for it lay in our language and language seemed to repeat it to us inexorably."

44. Domhoff, *Who Rules America?* argues that there is a "national upper class," the " 'American business aristocracy.' " While it "does not control every aspect of American political life," its influence is "sufficient to earn it the designation 'governing class' " (ibid., pp. 137, 156). The diversity and dispersion of activities by this class is, Domhoff holds, given unity by the common values, attitudes, and experiences of the class. It is interesting to note that Domhoff argues against the notion that "experts from the middle class have somehow displaced the American upper class as a governing class" (ibid., pp. 149-150). The experts are still chosen by the business aristocracy and take their orders accordingly.

45. Price, *Scientific Estate,* p. 17.

46. Ibid., p. 15.

47. Ibid., p. 36.

48. Ibid., p. 77.

49. Ibid., p. 55.

50. Ibid., p. 56.

51. Robert L. Heilbroner, *The Limits of American Capitalism* (New York: Harper & Row, 1966), pp. 65-134; Daniel Bell, "Notes on the Post-Industrial Society," in *The Technological Threat,* ed. Jack D. Douglas (Englewood Cliffs: Prentice-Hall, 1971), pp. 8-20. In Bell's words, p. 10, "To speak rashly: if the dominant figures of the past hundred years have been the entrepreneur, the businessman, and the industrial executive, the 'new men' are the scientists, the mathematicians, the economists, and the engineers of the new computer technology. And the dominant institutions of the new society—in the sense that they will provide the most creative challenges and enlist the richest talents—will be the intellectual institutions."

52. Price, *Scientific Estate,* p. 68.

53. Ibid., pp. 133-134.

54. Ibid., p. 133.

55. Ibid., p. 135.

56. Ibid.

57. Ibid., p. 137.

58. Bernard Crick, *The American Science of Politics* (Berkeley and Los Angeles: University of California Press, 1959), chaps. 11-12.

59. Price, *Scientific Estate,* pp. 147-148.

60. Ibid., pp. 148-149.

61. Ibid., pp. 204-207.

62. Price's mention of "faceless technocrats" appears in a quotation from Senator E. L. Bartlett of Alaska. Price summarizes the misapprehension here as follows: "But now that great issues turn on new scientific discoveries far too complicated for politicians to comprehend, many people doubt that representative institutions can still do their job. The fear that the new powers created by science may be beyond the control of constitutional processes, and that scientists may become a new governing clique or cabal of secret advisers, has begun to seem plausible" (ibid., p. 57). For a lively discussion of these matters from a British scholar, see Nigel Calder's *Technopolis: Social Control of the Uses of Science* (New York: Simon and Schuster, 1970).

63. Price, *Scientific Estate,* p. 153. Compare to *The Federalist*, No. 10, by James Madison in Alexander Hamilton, James Madison, and John Jay, *The Federalist Papers,* ed. Clinton Rossiter (New York: New American Library, 1961), pp. 77-84.

64. Price, *Scientific Estate,* chap. 3.

65. Ibid., p. 213.

66. Ibid., pp. 214-215.

67. Ibid., p. 121.

68. Ibid., p. 191.

69. See H. L. Nieburg, *In the Name of Science* (Chicago: Quadrangle, 1966); Melman, *Pentagon Capitalism;* Ralph E. Lapp, *The Weapons Culture* (Baltimore: Penguin Books, 1969); Richard Barnett, *The Economy of Death* (New York: Atheneum Publishers, 1969).

70. Nieburg, *Name of Science,* p. 198.

71. Ibid., p. 381.

72. Ibid., p. 187.

73. Price, *Scientific Estate,* p. 162.

74. Ibid.

75. Ibid., p. 45.

76. John Kenneth Galbraith, *The New Industrial State* (New York: New American Library, 1968), p. 82.

77. Ibid.

78. Ibid., chap. 2.

79. Ibid., p. 28.

80. Ibid., chap. 15.

81. Ibid., p. 315.

82. Ibid., p. 304.

83. Ibid., Chaps. 27, 29.

84. John Kenneth Galbraith, *Economics and the Public Purpose* (New York: New American Library, 1975).

85. Ibid., p. 405.

86. Ibid., pp. 400-401.

87. Galbraith, *Economics,* p. 215.

88. Galbraith, *New Industrial State,* p. 388.

89. Ibid., p. 392.

90. Ibid., p. 406.

91. Ibid., p. 301.

92. Ibid.

93. Ibid., p. 400.

94. Ibid., p. 406.

95. James Burnham, *The Managerial Revolution* (Bloomington: Indiana University Press, 1966), p. 101.

96. Galbraith, *Economics,* p. 292.

Notes to Chapter 5

1. Herbert Marcuse, *One-Dimensional Man: Studies in the Ideology of Advanced Industrial Society* (Boston: Beacon Press, 1964), p. 18.

2. See Sheldon S. Wolin's discussion in *Politics and Vision* (Boston: Little, Brown, 1960), p. 378; Henri de Saint-Simon, *Social Organization, the Science of Man and Other Writings,* ed. and trans. Felix Markham (New York, Harper & Row, 1964); Frank E. Manuel, *The New World of Henri Saint-Simon* (Notre Dame, Ind.: University of Notre Dame Press, 1963), chap. 27.

3. Writings that might be included in this tradition now include literally dozens of titles. The books I consider most important will be cited in specific footnotes. For the most complete bibliography of sources, consult Carl Mitcham and Robert Mackey, *Bibliography of the Philosophy of Technology* (Chicago: University of Chicago Press, 1973). It is not possible for me to acknowledge everything I have learned or been brought to think about through my readings of Ellul, Marcuse, Giedion, Mumford, Habermas, and others. If I have not seen farther, it is because there are giants standing on my shoulders.

4. A survey and critique of the penchant for naming the society is given in Henri Lefebvre's *The Sociology of Marx* (New York: Random House, 1968), pp. 192-197.

5. Examples of this tendency to use warnings as promotion can be found in Alvin Toffler, *Future Shock* (New York: Random House, 1970), and in Marshall McLuhan and Quentin Fiore, *The Medium Is the Massage* (New York: Bantam Books, 1967). For a while, the media and technology emphasis became an important theme in the public relations of many image-conscious American corporations. See Don Fabun's *Dynamics of Change* (Englewood Cliffs: Prentice-Hall, 1967), originally written as promotion for Kaiser Aluminum.

6. Some interesting examples of artistic critiques of technological society are to be found in popular recordings: Captain Beefheart and His Magic Band, *Trout Mask Replica,* Straight STS 1053; The Mothers of Invention, *We're Only in It for the Money,* Verve V6 5045X; and The Firesign Theatre, *Don't Crush That Dwarf, Hand Me the Pliers,* Columbia C30102, and *I Think We're All Bozos on This Bus,* C30737.

7. Jacques Ellul, *The Technological Society,* trans. John Wilkinson (New York: Alfred A. Knopf, 1964), pp. 173-177, 168-169.

8. See, for example, Karl Marx, *The Civil War in France* (New York: International Publishers, 1940), and Karl Marx and Friedrich Engels, *Manifesto of the Communist Party* in Marx and Engels, *Basic Writings on Politics and Philosophy,* ed. Lewis S. Feuer (New York: Doubleday & Company, Anchor Books, 1959), pp. 1-41.

9. See Karl Jaspers, *Man in the Modern Age,* trans. Eden and Cedar Paul (Garden City: Doubleday & Company. Anchor Press, 1957), pp. 21-22.

10. Ellul, *Technological Society,* p. 238.

11. Jean Piaget identifies "self-regulation" as a basic property of all structures, "self-regulation entailing self-maintenance and closure," in his *Structuralism,* trans. and ed. Chaninah Maschler (New York: Basic Books, 1970), pp. 13-16.

12. Jean-Jacques Rousseau, *The First and Second Discourses,* ed. Roger D. Masters and trans. Roger D. and Judith R. Masters (New York: St. Martin's Press, 1964), pp. 60-62, 151-155.

13. Michael Oakeshott, *Rationalism in Politics* (New York: Basic Books, 1962), pp. 10-11, 80-110.

14. A particularly good discussion of this meaning of rationality is found in Hubert L. Dreyfus, *What Computers Can't Do* (New York: Harper & Row, 1972), pp. xv-xxxv.

15. Ellul, *Technological Society,* p. 79.

16. Max Weber, *From Max Weber,* trans. and ed. H. H. Gerth and C. Wright Mills (New York: Oxford University Press, Galaxy Books, 1958), pp. 155, 293. See also Herbert Marcuse's "Industrialization and Capitalism in the Work of Max Weber," in his *Negations: Essays in Critical Theory* (Boston: Beacon Press, 1968), pp. 201-226, and Jurgen Habermas, *Toward a Rational Society,* trans. Jeremy Shapiro (Boston: Beacon Press, 1970), pp. 50-122.

17. David S. Landes, *The Unbound Prometheus: Technological Change and Industrial Development in Western Europe from 1750 to the Present* (Cambridge: At the University Press, 1969), pp. 21-26, 546.

18. James D. Thompson, *Organizations in Action* (New York: McGraw-Hill, 1967), p. 14. "Technical rationality," he argues, "can be evaluated by two criteria: instrumental and economic. The essence of the instrumental question is whether the specified actions do in fact produce the desired outcome, and the instrumentally perfect technology is one which inevitably achieves such results."

19. Ellul's definition of technique stresses this conclusion. "In our technological society, *technique is the totality of methods rationally arrived at and having absolute efficiency* (for a given stage of development) in *every* field of human activity." Ellul, *Technological Society,* p. xxv.

20. Lewis Mumford, *The Myth of the Machine: Technics and Human Development* (New York: Harcourt, Brace and World, 1967), chap. 11.

21. See John Blair, *Economic Concentration: Structure, Behavior and Public Policy* (New York: Harcourt Brace Jovanovich, 1972).

22. Lewis Mumford, *Art and Technics* (New York: Columbia University Press, 1960).

23. Francis Bacon, *Novum Organum,* in *Selected Writings of Francis Bacon,* ed. Hugh G. Dick (New York: The Modern Library, 1955), p. 534.

24. Siegfried Giedion, *Mechanization Takes Command* (New York: W. W. Norton, 1969), pp. 32-33. See also Reyner Banham's excellent discussion of technology and modern architecture, *Theory and Design in the First Machine Age* (New York: Praeger Publishers, 1967).

25. Giedion ties together some of the notions we are developing in the following statement in his conclusion, "Man in Equipoise." "From the very first it was clear that mechanization involved a division of labor. The worker cannot manu-

facture a product from start to finish; from the standpoint of the consumer the product becomes increasingly difficult to master. When the motor of his car fails, the owner often does not know which part is causing the trouble; an elevator strike can paralyze the whole life of New York. As a result, the individual becomes increasingly dependent on production and on society as a whole, and relations are far more complex and interlocked than in any earlier society. This is one reason why today man is overpowered by means." Giedion, *Mechanization,* p. 714.

26. Thorstein Veblen, *The Engineers and the Price System* (New York: The Viking Press, 1954), p. 52.

27. Ibid., p. 122.

28. Ibid., p. 57.

29. Ellul, *Technological Society,* p. 193.

30. Herbert Marcuse, *Eros and Civilization* (New York: Random House, Vintage Books, 1962), p. vii.

31. Ibid.

32. G. W. F. Hegel, *The Phenomenology of Mind,* trans. J. B. Baillie (London: George Allen & Unwin Ltd., 1931), pp. 236-237.

33. Ibid., p. 238.

34. Friedrich Nietzsche, *The Genealogy of Morals,* with *The Birth of Tragedy,* trans. Francis Golffing (New York: Doubleday & Company, Anchor Books, 1956). A similar argument with "truth" and "falsity" as the focus is found in Nietzsche's essay, "Truth and Falsity in an Ultramoral Sense," in *The Philosophy of Nietzsche,* ed. Geoffrey Clive (New York: New American Library, Mentor Books, 1965), pp. 503-515.

35. *The Oxford English Dictionary,* "robot" (Oxford: Clarendon Press, 1971).

36. Karel Čapek, *R. U. R. (Rossum's Universal Robots),* in *Of Men and Machines,* ed. Arthur O. Lewis, Jr. (New York: E. P. Dutton, 1963), p. 52.

37. See Hiram Haydn's *The Counter-Renaissance* (New York: Harcourt, Brace & World, 1950), pp. 295-296, and Michael Walzer, *The Revolution of the Saints* (London: Weidenfeld and Nicoloson, 1966), chap. 1. Both works give an interesting comparison of medieval and modern notions of order and membership.

38. Thomas Carlyle, *Sartor Resartus,* in Carlyle's Complete Works, The Vellum Edition, Vol. I (Boston: Dana Estes and Charles E. Lauriat, 1884), pp. 31-32.

39. E. J. Dijksterhuis, *The Mechanization of the World Picture* trans. C. Dikshoorn (Oxford: Oxford University Press, 1961).

40. Julien Offray de La Mettrie, *Man a Machine* (La Salle, Ill.: Open Court Publishing Company, 1961), p. 93.

41. For the major arguments now marshalled in this continuing debate, see Dean

E. Woolridge, *Mechanical Man: The Physical Basis of Intelligent Life* (New York: McGraw-Hill, 1968); Floyd W. Matson, *The Broken Image: Man, Science and Society* (New York: Doubleday & Company, Anchor Books, 1966); B. F. Skinner, *Beyond Freedom and Dignity* (New York: Alfred A. Knopf, 1971), p. 204. Woolridge, p. 204, explains in defense of the mechanistic view that "men who know they are machines should be able to bring a higher degree of objectivity to bear on their problems than machines that think they are Men."

42. See Leo Marx, *The Machine in the Garden: Technology and the Pastoral Ideal in America* (New York, Oxford University Press, 1964).

43. Lewis Mumford, *The Myth of the Machine: The Pentagon of Power* (New York: Harcourt Brace Jovanovich, 1970), pp. 163-169.

44. Rousseau, *First and Second Discourses,* p. 151.

45. Ibid., p. 152.

46. Ibid., pp. 153-154.

47. Ralph Waldo Emerson, *Works and Days,* in *Of Men and Machines,* ed. Arthur O. Lewis (New York: E. P. Dutton, 1963), p. 68.

48. "Civilization is a reagent and eats away the old traits." Emerson in *English Traits,* cited by *Oxford English Dictionary,* "reagent."

49. Emerson, *Works and Days,* p. 68.

50. Bruno Bettelheim, *The Informed Heart: Autonomy in a Mass Age* (New York: The Free Press, 1960), p. 48.

51. Ibid., p. 49.

52. Ibid.

53. Thorstein Veblen, *The Theory of Business Enterprise* (New York: New American Library, Mentor Books, 1970), p. 144.

54. Ibid., p. 146.

55. Ibid., pp. 146-147.

56. Ibid., p. 177.

57. "Without exception in the course of history, *technique belonged to a civilization* and was merely a single element among a host of nontechnical activities. Today *technique has taken over the whole of civilization.* Certainly, technique is no longer the simple machine substitute for human labor. It has come to be the 'intervention into the very substance not only of the inorganic but also of the organic.' " Ellul, *Technological Society,* p. 128.

58. Ibid., p. 98.

59. Emerson, *Works and Days,* p. 64.

60. Marcuse, *One-Dimensional Man,* p. 33.

61. Ibid., pp. 36-37.

62. Ibid., p. 37.

63. Ibid., p. 42.

64. Friedrich Georg Juenger, *The Failure of Technology* (Chicago: Henry Regnery, 1956), p. 8.

65. Hannah Arendt, *The Human Condition* (Chicago: University of Chicago Press, 1958), p. 131.

66. Jules Henry, *Culture Against Man* (New York: Random House, Vintage Books, 1965), pp. 15-24.

67. Marcuse, *Eros and Civilization,* pp. 32-34, and *One-Dimensional Man,* chaps. 1-3.

68. Ellul's assessment of Marcuse's position in the debate is offered in his *Autopsy of Revolution,* trans. Patricia Wolf (New York: Alfred A. Knopf, 1971), pp. 287-290. One of the kinder things he says is the following: "Although Freud's work suggests a revolutionary approach to sexual repression, which begins in the family and is perpetuated by a network of social relationships, he was cautious and never indulged in the acrobatic fantasies of Marcuse, not because he was hopelessly bourgeois, but out of recognition of the unreliability of the unconscious, which made repression necessary, and out of a somewhat skeptical view of revolution. He thought that revolution could not 'change life,' as the reinforced patterns of servility, guilt, and repression would be likely to reappear in seemingly different social surroundings. I accept the logic of that view, whereas 'Freudian-Marxist syntheses' strike me as so much haphazard verbiage—but dangerous, still, as all meaningless verbiage is, for they shunt the revolutionary impulse into dead storage, identifying the sexual explosion with revolution, and giving sterile and brutish expression to the whole legacy of revolution" (ibid., p. 287). To the best of my knowledge, Herbert Marcuse has not yet published a response to the work of Jacques Ellul.

69. Ellul, *Technological Society,* p. 64-79.

70. Ibid., p. 65.

71. Ibid., p. 191.

72. Ibid.

73. Ibid., p. 192.

74. Ibid.

75. Lewis Mumford, *Technics and Civilization* (New York: Harcourt, Brace & World, 1934), p. 232.

76. Ellul, *Technological Society,* p. 142.

77. Ibid., p. 325. "The milieu in which he lives is no longer his. He must adapt himself, as though the world were new, to a universe for which he was not created."

78. The best attempt is the provocative essay *The Dehumanization of Art* by José Ortega y Gasset (Princeton: Princeton University Press, 1951).

79. Giedion, *Mechanization,* p. vi.

80. Ibid., p. 97.

81. Ibid., p. 533.

82. Ibid., p. 246.

83. Ibid., p. 712.

84. Ellul, *Technological Society,* p. 183.

85. Ibid., p. 193.

86. Ibid., p. 168.

87. Ibid., pp. 168-169.

88. Ibid., p. 325.

89. Ibid., chap. 5.

90. Ibid., p. 395.

91. Ibid.

92. Jaspers, *Man,* p. 73.

93. Ellul, *Technological Society,* p. 219.

94. Ibid.

95. Michael Rose, *Computers, Managers and Society* (Baltimore: Penguin Books, 1969), p. 159.

96. Ibid., p. 161.

97. See Samuel C. Florman, *The Existential Pleasures of Engineering* (New York: St. Martin's Press, 1976).

98. An eloquent analysis of this kind is given in Percy Bysshe Shelley's "A Defense of Poetry," in *The Selected Poetry and Prose of Shelley,* ed. Harold Bloom (New York: New American Library, 1966), pp. 415-448. Shelley observes that the poets, once the true "legislators or prophets" of the world, "have been challenged to resign the civic crown to reasoners and mechanists" (ibid., pp. 419, 439). He predicts that a withering of the poetic faculty can only spell disaster for a society driven by a new and highly productive rational knowledge. "We want the creative faculty to imagine that which we know; we want the generous impulse to act that which we imagine; we want the poetry of life: our calculations have outrun conception; we have eaten more than we can digest" (ibid., p. 441). Shelley sees very clearly the conflict of the two worlds of complexity. He also anticipates with considerable accuracy a problem that would arise in the twentieth century, information overload. Compare to the issues discussed in chapter 7 of this book.

99. See Peter Gay, *Weimar Culture: The Outsider as Insider* (New York: Harper & Row, 1968); Hans Kohn, *The Mind of Germany* (New York: Harper & Row, 1965).

100. Marcuse, *One-Dimensional Man*, p. 31.

101. Ibid., p. 14.

102. Ibid., p. 103.

103. Ibid., p. 104.

104. W. T. Singleton, *Man-Machine Systems* (Harmondsworth, England: Penguin Books, 1974), p. 37.

105. For an interesting treatment of systems analysis in this context, see Ida R. Hoos, *Systems Analysis in Public Policy: A Critique* (Berkeley and Los Angeles: University of California Press, 1972).

106. See Paul Brett Hammond, "Language, Social Choice and Systems Analysis in Theory and Practice," 1975, an unpublished paper.

107. I confess that this sentence is my own. It would, however, probably not be noticed as anything out of the ordinary in the conversations of today's engineers, planners, and social scientists.

108. See Marcuse, *Negations*, pp. 201-226.

109. Emmanuel G. Mesthene, *Technological Change: Its Impact on Man and Society* (New York: New American Library, 1970), pp. 15-20.

110. Ellul, *Technological Society*, p. 220.

111. I am not saying that the notion of "use" with regard to advanced technology is totally nonsensical, only that it is a misleading concept in many cases. Most talk about "using" large-scale, complex technical systems is, I believe, merely a bad linguistic habit, an anachronism that leaves the speaker impotent in dealing with the problem at hand.

112. Kenneth Keniston, *The Uncommitted: Alienated Youth in American Society* (New York: Delta, 1967), p. 379.

113. Ibid., p. 366.

114. Ellul, *Technological Society*, p. 277.

115. Jacques Ellul, *Presence of the Kingdom*, trans. Olive Wyon (New York: Seabury Press, 1967), p. 66.

116. Ibid.

117. Gene L. Maeroff, "Writing Test for College is Urged," *New York Times*, January 25, 1976. The statement is by Albert G. Sims, a vice-president of the College Entrance Examination Board.

118. Ellul, *Technological Society*, p. 80

119. Jacob Schmookler, *Invention and Economic Growth* (Cambridge, Mass.:

Harvard University Press, 1966). See also *The Economics of Technological Change*, ed. Nathan Rosenberg (Harmondsworth, England: Penguin Books, 1971), a good anthology of essays containing works by Schmookler, Joseph Schumpeter, A. P. Usher, Robert Solow, and other economists.

Notes to Chapter 6

1. Jacques Ellul, *The Technological Society*, trans. John Wilkinson (New York: Alfred A. Knopf, 1964), p. 166.

2. Ibid., p. 184.

3. Ibid., p. 177.

4. A sample of this response is found in Melvin Kranzberg, "Historical Aspects of Technology Assessment," *Technology Assessment Hearings Before the Subcommittee on Science, Research and Development of the U.S. House of Representatives* (Washington: U.S. Government Printing Office, 1970). After a brief survey of the ideas of Ellul, Mumford, and Marcuse, Kranzberg concludes: "While such wholesale indictments may stimulate nihilistic revolutionary movements, they really tell us very little about what can be done to guide and direct technological innovation along socially beneficial lines" (ibid., p. 385).

5. Another interesting side to the "forward"-"backward" view counsels that the forward direction is ineluctable. In his statement to a U.S. Senate hearing in 1970, Harvey Brooks took care to deny the proposition that "technological progress" is a "largely autonomous development." But he went on to say, "While this pessimistic view of technology is not without evidence to support it, I believe it represents only a partial truth. Furthermore, it is an essentially sentimental and irrational view, because man in fact has *no choice* but to push forward with his technology. The world is already *irrevocably committed to a technological culture* [emphasis added]." Reprinted ibid., p. 331.

6. Herbert Marcuse, *Negations: Essays in Critical Theory*, trans. Jeremy J. Shapiro (Boston: Beacon Press, 1969), p. xiii.

7. John Kenneth Galbraith, *The New Industrial State* (New York: The New American Library, 1968), p. 37.

8. Ibid., p. 39.

9. Ibid.

10. Ibid., p. 41.

11. Compare Grant McConnell, *Private Power and American Democracy* (New York: Alfred A. Knopf, 1966), and the reports of the Ralph Nader Study groups: James S. Turner, *The Chemical Feast: The Ralph Nader Study Group Report on the Food and Drug Administration* (New York: Grossman Publishers, 1970); Robert Fellmeth, *The Interstate Commerce Omission: The Ralph Nader Study Group Report on the Interstate Commerce Commission and Transportation* (New York: Grossman Publishers, 1970).

12. Gene Marine and Judy Van Allen, *Food Pollution* (New York: Holt, Rinehart & Winston, 1972).

13. See Clark R. Mollenhoff, *The Pentagon: Politics, Profits and Plunder* (New York: Pinnacle Books, 1972); Murray Weidenbaum, "Arms and the American Economy: A Domestic Convergence Hypothesis," *Quarterly Review of Economics and Business* 8 (Spring 1968); Ralph Lapp, *The Weapons Culture* (Baltimore: Penguin Books, 1968); Seymour Melman, *Pentagon Capitalism: The Political Economy of War* (New York: McGraw-Hill, 1970).

14. Ellul, *Technological Society,* p. 225.

15. Ibid., p. 221.

16. See Theodor Adorno, *Minima Moralia,* trans. E. F. N. Jephcott (London: NLB, 1974); Max Horkheimer, *Critical Theory,* trans. Matthew J. O'Connell et al. (New York: Herder and Herder, 1972); Jürgen Habermas, *Legitimation Crisis,* trans. Thomas McCarthy (Boston: Beacon Press, 1975). An interesting, polemical review of the progress of the Frankfurt school is given in Göran Therborn's article, "A Critique of the Frankfurt School," *New Left Review,* no. 63 (September-October 1970): 65-96.

17. "Propaganda is a set of methods employed by an organized group that wants to bring about the active or passive participation in its actions of a mass of individuals, psychologically unified through psychological manipulations and incorporated in an organization." Jacques Ellul, *Propaganda,* trans. Konrad Kellen (New York: Alfred A. Knopf, 1967), p. 61.

18. See John Blair, *Economic Concentration: Structure, Behavior and Public Policy* (New York: Harcourt Brace Jovanovich, 1972), chap. 5.

19. Lewis Mumford, *Technics and Civilization* (New York: Harcourt Brace & World, 1934), pp. 223-224.

20. Ellul, *Technological Society,* p. 237.

21. Lewis Mumford, *The Myth of the Machine: The Pentagon of Power* (New York: Harcourt, Brace Jovanovich, 1970), pp. 239, 268-273.

22. Ellul, *Technological Society,* pp. 193-194.

23. F. J. Roethlisberger and William J. Dickson, *Management and the Worker* (Cambridge, Mass.: Harvard University Press, 1939); Elton Mayo, *The Social Problems of Industrial Civilization* (Boston: Graduate School of Business Administration, Harvard University, 1945).

24. Richard M. Cyert and James G. March, *A Behavioral Theory of the Firm* (Englewood Cliffs: Prentice-Hall, 1963).

25. See Arthur L. Stinchcombe, "Bureaucratic and Craft Administration of Production," *Administrative Science Quarterly* 4 (1959): 168-187; Peter M. Blau and W. Richard Scott, *Formal Organizations: A Comparative Approach* (San Francisco: Chandler Publishing Co., 1962), pp. 207-214.

26. See Warren G. Bennis, *Changing Organizations* (New York: McGraw-Hill, 1963); Frederick C. Thayer, *An End to Hierarchy! An End to Competition!* (New York: New Viewpoints, 1973). The research and thinking of political scientist Robert Biller, none of it yet published, focus upon the rise of "collegial" forms of organization in highly sophisticated technical fields.

27. David Braybrooke and Charles E. Lindblom, *A Strategy of Decision* (New York: The Free Press, 1970).

28. Warren G. Bennis and Philip E. Slater, *The Temporary Society* (New York: The Free Press, 1970).

29. Ellul, *Technological Society,* p. 307.

30. Ibid., p. 196.

31. Ibid., p. 307.

32. Ibid., p. 274.

33. Ibid.

34. Ibid., p. 259.

35. Ibid., p. 186.

36. See Jean Meynaud, *Technocracy,* Trans. Paul Barnes (New York: The Free Press, 1964), p. 206, for an agonizing effort to appraise the situation. "The spirit of our day," he observes, "directed towards the search for maximal productivity, is definitely favourable to technocratic ideology. At this point, it is essential to identify those who will benefit from the movement."

37. Herbert Marcuse, *Negations: Essays in Critical Theory* (Boston: Beacon Press, 1968), p. 224.

38. Herbert Marcuse, *One-Dimensional Man: Studies in the Ideology of Advanced Industrial Society* (Boston: Beacon Press, 1964), p. 32.

39. Karl Marx and Friedrich Engels, *Manifesto of the Communist Party,* in *Marx and Engels: Basic Writings on Politics and Philosophy,* ed. Lewis Feuer (New York: Doubleday & Company, Anchor Books, 1959), p. 28.

40. V. I. Lenin, "The Tasks of the Youth Leagues," speech delivered at the Third All-Russia Congress of the Russian Young Communist League, October 2, 1920, in *Lenin: Selected Works* (New York: International Publishers, 1971), p. 607.

41. Ibid., p. 612.

42. V. I. Lenin, "The Immediate Tasks of the Soviet Government," in *Lenin,* p. 417.

43. Ibid. In his essay " 'Left-Wing' Childishness and Petty–Bourgeois Mentality," Lenin says of the Menshevik critics of scientific management: "It would be extremely useful indeed for the workers to think over the reason why such

lackeys of the bourgeoisie should incite the workers to resist the Taylor system and the 'establishment of trusts' " (ibid., p. 450).

44. The essay was left unfinished. "It is more pleasant and useful to go through the 'experience of revolution' than to write about it," Lenin observes. *State and Revolution* (New York: International Publishers, 1943), p. 101.

45. Ibid., pp. 7-10.

46. Ibid., p. 16.

47. Ibid., pp. 15-20.

48. Ibid., p. 27.

49. Ibid., p. 80.

50. Ibid., p. 84.

51. Ibid., p. 82.

52. Ibid.

53. Ibid.

54. "Overthrow the capitalists, crush with the iron hand of the armed workers the resistance of these exploiters, break the bureaucratic machine of the modern state—and you have before you a mechanism of the highest technical equipment, freed of 'parasites,' capable of being set into motion by the united workers themselves who hire their own technicians, managers, bookkeepers, and pay them *all,* as, indeed, every 'state' official, with the usual workers' wage." Ibid., p. 43.

55. Ibid., p. 40.

56. Ibid., p. 52.

57. Ibid.

58. Ibid., p. 91.

59. Ibid., pp. 91-92.

60. Ibid., p. 92.

61. Ibid., p. 83.

62. Ibid.

63. See, for example, Rosa Luxembourg, "The Russian Revolution," in *Rosa Luxembourg Speaks,* ed. Mary Alice Waters (New York: Pathfinder Press, 1970), pp. 367-395; E. H. Carr, *A History of Soviet Russia* (Middlesex, England: 1968); Isaac Deutscher, *Stalin: A Political Biography* (New York: Oxford University Press, 1967); Daniel Cohn-Bendit and Gabriel Cohn-Bendit, *Obsolete Communism, The Left-Wing Alternative,* trans. Arnold Pomerans (New York: McGraw-Hill, 1968).

64. Deutscher, *Stalin,* especially chap. 8, "The Great Change."

65. Lenin, "Tasks of the Youth Leagues," p. 611.

66. Karl Marx, quoted in Lenin, *State and Revolution,* pp. 78-79.

67. Lenin, *State and Revolution,* p. 79.

68. Ibid.

69. Ibid.

70. Ibid., p. 83.

71. Ibid., pp. 83-84.

72. See *China: Science Walks on Two Legs* (New York: Avon Books, 1974); E. L. Wheelwright and Bruce McFarlane, *The Chinese Road to Socialism: Economics of the Cultural Revolution* (New York: Monthly Review Press, 1970). For a critique of exorbitant left-wing hopes for Chinese socialism, see Gilbert Padoul, "China 1974: Problems Not Models," *New Left Review,* no. 89 (January-February 1975): 73-84.

73. Jürgen Habermas, *Toward a Rational Society,* trans. Jeremy Shapiro (Boston: Beacon Press, 1970), chap. 6.

74. Ellul, *Technological Society,* p. 291.

75. Ibid., p. 282.

Notes to Chapter 7

1. Thomas Hobbes, quoted in Richard Peters, *Hobbes* (Baltimore: Penguin Books, 1967), p. 68.

2. Plato, *The Republic,* I. 332-343, VI. 488-489.

3. Thomas Hobbes, *Leviathan,* ed. Michael Oakeshott (New York: Collier, 1962), p. 19.

4. Jacques Ellul, *The Technological Society,* trans. John Wilkinson (New York: Alfred A. Knopf, 1964), p. 134.

5. Ibid.

6. Ibid.

7. Ibid.

8. Arthur Koestler, *The Act of Creation* (New York: Dell Publishing, 1967), p. 264.

9. See my "Complexity and the Limits of Human Understanding," in Todd R. La Porte, ed., *Organized Social Complexity: Challenge to Politics and Policy* (Princeton: Princeton University Press, 1975), pp. 40-76.

10. Ellul, *Technological Society,* pp. 141-146.

11. See note 9.

12. Amitai Etzioni, *The Active Society: A Theory of Societal and Political Processes* (New York: The Free Press, 1968), pp. 142-146.

13. Ludwig von Bertalanffy, *General System Theory: Foundations, Development, Applications* (New York: George Braziller, 1968), p. 30.

14. Ibid., p. 33.

15. Ibid., pp. vii-viii.

16. Paul Valéry, "Unpredictability," in his *History and Politics,* trans. Denise Folliot and Jackson Mathew (New York: Pantheon Books, 1962), p. 69.

17. H. G. Wells, *The Mind at the End of Its Tether* (London: William Heinemann Ltd., 1945), p. 4.

18. Ibid., p. 34.

19. David Braybrooke and Charles Lindblom, *A Strategy of Decision* (New York: The Free Press, 1963). Another interesting interpretation of the political order as a self-adjusting mechanism is the cybernetic theory of Karl W. Deutsch, *Nerves of Government: Models of Political Communication and Control* (New York: The Free Press, 1966). This writer's critique of Deutsch's theory is given in "Cybernetics and Political Language: A Response to Karl Deutsch," *Berkeley Journal of Sociology* 14 (1969): 1-17.

20. Charles E. Lindblom, "The Science of Muddling Through," *Public Administration Review* 19 (Spring 1959): 80.

21. Braybrooke and Lindblom, *Strategy of Decision,* p. 85.

22. Lindblom, "Science of Muddling Through," p. 85.

23. Ibid.

24. Ibid.

25. Charles E. Lindblom, *The Intelligence of Democracy* (New York: The Free Press, 1965).

26. Adam Smith, *An Inquiry into the Nature and Causes of the Wealth of Nations* (New York: The Modern Library), p. 423.

27. Braybrooke and Lindblom, *Strategy of Decision,* p. 73.

28. Paul Goodman, *New Reformation: Notes of a Neolithic Conservative* (New York: Random House, 1970), pp. 192-193.

29. Jacques Ellul's discussion of "the necessary and the ephemeral" is useful in this regard. *The Political Illusion,* trans. Konrad Kellen (New York: Alfred A. Knopf, 1967), chap. 1.

30. See Joseph Weizenbaum, *Computer Power and Human Reason: From Judgment to Calculation* (San Francisco: W. H. Freeman and Company, 1976), chap. 9.

31. "Vietnam Bombing Evaluation by Institute for Defense Analyses," in *The Pentagon Papers* (New York: Bantam Books, 1971), p. 502.

32. Ibid., p. 505.

33. Ibid.

34. Ibid., p. 506.

35. Ibid., pp. 506-507.

36. Ibid., pp. 507-508.

37. Harold L. Wilensky, *Organizational Intelligence: Knowledge and Policy in Government and Industry* (New York: Basic Books, 1967), p. 48.

38. Ibid., p. 42.

39. Ibid.

40. A discussion of the electronic battlefield and of other aspects of advanced technological warfare is contained in Michael T. Klare's *War Without End* (New York: Random House, Vintage Books, 1972), chaps. 5-7.

41. Barbara Tuchman, *The Guns of August* (New York: Dell Publishing Co., 1963), pp. 94-95.

42. Ibid., p. 99.

43. See David Kraslow and Stuart H. Loory, *The Secret Search for Peace in Vietnam* (New York: Random House, Vintage Books, 1968), pp. 3-5.

44. *Report to the President and the Secretary of Defense on the Department of Defense by the Blue Ribbon Defense Panel* (Washington, D.C.: Government Printing Office, 1970).

45. As reported by Hannah Arendt, *Eichmann in Jerusalem: A Report on the Banality of Evil*, rev. and enl. (New York: The Viking Press, 1964), p. 57.

46. Ibid., p. 25.

47. Raphael Littauer and Norman Uphoff, eds., *The Air War in Indo-China* (Boston: Beacon Press, 1972), pp. 158-159.

48. Ibid., p. 159.

49. Marvin Minsky, "Steps Toward Artificial Intelligence," in *Computers and Thought*, ed. Edward A. Feigenbaum and Julian Feldman (New York: McGraw-Hill, 1963), p. 447. I am indebted to Joseph Weizenbaum for calling this point to my attention.

50. Max Horkheimer, *Eclipse of Reason* (New York: Seabury Press, 1974), p. 102.

Notes to Chapter 8

1. Mary Shelley, *Frankenstein, or The Modern Prometheus*, in *Three Gothic Novels*, ed. Peter Fairclough (Harmondsworth: Penguin Books, 1968), p. 295. The Penguin edition reprints the text of *Frankenstein* published in London in 1831 by Colburn and Bentley and contains Mary Shelley's final revisions. Compare to the first edition in *Frankenstein, or The Modern Prometheus* (the 1818 text), ed. James Rieger (New York: Bobbs-Merrill, 1974). Mary Shelley's life and writings are discussed in Eileen Bigland, *Mary Shelley* (London: Cassell, 1959); Elizabeth Nitchie, *Mary Shelley, Author of "Frankenstein"* (New Brunswick:

Rutgers University Press, 1953); Margaret (Carter) Leighton, *Shelley's Mary: The Life of Mary Godwin Shelley* (New York: Farrar, Straus, & Giroux, 1973). Those interested in a quick, provocative account of the story behind the novel can turn to Samuel Rosenberg's essay, "Frankenstein, or Daddy's Little Monster," in *The Confessions of a Trivialist* (Baltimore: Penguin Books, 1972).

2. Shelley, *Frankenstein,* p. 307.

3. Ibid., p. 312.

4. Ibid., pp. 318-319.

5. Ibid., p. 319.

6. Ibid., p. 364.

7. Ibid., p. 363.

8. Ibid., p. 364.

9. Ibid., p. 366.

10. Ibid., p. 413.

11. Ibid., p. 435.

12. Ibid., pp. 435-436.

13. Ibid., p. 436.

14. Ibid., p. 439.

15. See Rosenberg, "Frankenstein." There is now a growing literature on the historical background of the novel. See N. H. Brailsford, *Shelley, Godwin and Their Circle* (New York: Henry Holt and Co., 1913); Christopher Small, *Mary Shelley's Frankenstein—Tracing the Myth* (Pittsburgh: University of Pittsburgh Press, 1973); Radu Florescu, *In Search of Frankenstein* (Boston: New York Graphic Society, 1975); Ellen Moers, "Female Gothic: The Monster's Mother," *New York Review of Books,* March 21, 1974.

16. See Mario Praz's introduction to *Three Gothic Novels,* pp. 25-31.

17. Percy Bysshe Shelley, *Prometheus Unbound,* in *The Selected Poetry and Prose of Shelley,* ed. Harold Bloom (New York: New American Library, 1966), p. 121.

18. Ibid., p. 124.

19. Spiro T. Agnew, "Address by the Vice President of the United States to the Printing Industries of America Convention," transcript, New York, New York, July 12, 1972. See also Mr. Agnew's philosophy of progress in "Address by the Vice President of the United States at the Alaska Republican Luncheon," transcript, Fairbanks, Alaska, July 24, 1972.

20. The French philosophes, for example, saw progress as the development of education and moral virtue, as well as the growth of science and technology. They appreciated the contributions of the past to this development, rejecting the now

common view that only the latest thing counts. In the *Encyclopedia,* Diderot summarizes this outlook: "The aim of an *encyclopedia* is to collect all the knowledge scattered over the face of the earth, to present its general outlines and structure to the men with whom we live, and to transmit this to those who will come after us, so that the work of past centuries may be useful to the following centuries, that our children, by becoming more educated, may at the same time become more virtuous and happier, and that we may not die without having deserved well of the human race." *The Encyclopedia, Selections,* ed. and trans. Stephen J. Gendzier (New York: Harper & Row, 1967), p. 92.

21. Representative works in this genre are: *Technology: Processes of Assessment and Choice,* Report of the National Academy of Sciences, Committee on Science and Astronautics, U.S. House of Representatives (Washington, D.C., Government Printing Office, 1969); Herman Kahn and Anthony J. Wiener, *The Year 2000: A Framework for Speculation on the Next Thirty-Three Years* (New York: Macmillan, 1967); *Harvard University Program on Technology and Society, 1964-1972: A Final Review* (Cambridge, Mass.: Harvard University Press, 1972). Other prominent voices in the conversation include the followers of R. Buckminster Fuller, for example, John McHale, *The Future of the Future* (New York: George Braziller, Inc., 1969), and the ubiquitous environmentalists, for example, Barry Commoner, *The Closing Circle: Nature, Man and Technology* (New York: Alfred A. Knopf, 1971).

22. See *The Impacts of Snow Enhancement: Technology of Winter Orographic Snowpack Augmentation in the Upper Colorado River Basin,* comp. Leo W. Weisbecker (Norman, Okla.: University of Oklahoma Press, 1974).

23. Under the National Environmental Policy Act of 1970, all government agencies doing work likely to have an impact on the environment are required to prepare a detailed study covering the likely effects of their projects on the environment. These impact statements are supposed to consider alternative plans and to show why the one chosen is preferable. The law has given birth to a small new industry of impact statement writing, much of which operates under the influence of the science of public relations.

24. For the distinction between "Technologies and Supporting Systems," see *Technology,* pp. 15-18.

25. See, for example, Charles Reich's *The Greening of America* (New York: Random House, 1970). One began to wonder about the viability of Reich's "Consciousness III" when one noticed how well it fit with the operation of the modern corporation.

26. Stanley Cavell, "The Avoidance of Love," in his *Must We Mean What We Say* (New York: Charles Scribner's Sons, 1969), pp. 337-353; Nadezhda Mandelstam, *Hope Abandoned: A Memoir,* trans. Max Hayward (London: Collins and Harvill Press, 1974), p. 182.

27. See Hannah Arendt, *The Human Condition* (Chicago: University of Chicago

Press, 1958), and chap. 6 of *On Revolution* (New York: The Viking Press, 1963); Sheldon Wolin, *Politics and Vision* (Boston: Little, Brown, 1960); Carole Pateman, *Participation and Democratic Theory* (Cambridge: Cambridge University Press, 1970).

28. See Rosabeth Moss Kanter, *Commitment and Community: Communes and Utopias in Sociological Perspective* (Cambridge, Mass.: Harvard University Press, 1972); Sam Dolgoff, ed., *The Anarchist Collectives: Workers' Self-Management in the Spanish Revolution, 1936-1938* (New York: Free Life Editions, 1974); Albert Soboul, *The Sans-Culottes: The Popular Movement and Revolutionary Government, 1793-1794,* trans. Remy Inglish Hall (New York: Doubleday & Company, Anchor Books, 1972); Stewart Edwards, *The Paris Commune 1871* (Chicago: Quadrangle Books, 1971).

29. See Terrence E. Cook and Patrick M. Morgan, eds. *Participatory Democracy* (San Francisco: Canfield Press, 1971).

30. Paul Goodman, *People or Personnel and Like a Conquered Province* (New York: Random House, Vintage Books, 1968), pp. 297-316; Murray Bookchin, *Post-Scarcity Anarchism* (Berkeley: Ramparts Books, 1971); Herbert Marcuse, *An Essay on Liberation* (Boston: Beacon Press, 1969); Jacques Ellul, *Autopsy of Revolution,* trans. Patricia Wolf (New York: Alfred A. Knopf, 1971), chap. 5.

31. Erich Fromm finds in this tendency the foremost principle of action in the technological society: "something ought to be done because it is technically possible to do so." *The Revolution of Hope* (New York: Bantam Books, 1968), p. 33. In some understandings this principle or the motive it expresses is taken to be "the technological imperative" itself. I have not adopted that definition here, preferring to employ the term in the context presented in chapters 2 and 6. The phenomenon Fromm and others have noticed is perhaps best called "techno-mania."

32. Gertrude Stein, *Look At Me Now and Here I Am, Writings and Lectures 1909-45,* ed. Patricia Meyerowitz (Baltimore: Penguin Books, 1971), p. 58.

33. See E. F. Schumacher, *Small is Beautiful: Economics As If People Mattered* (New York: Harper & Row, 1973); Ivan Illich, *Tools for Conviviality* (New York: Harper & Row, 1973); Wilson Clark, *Energy for Survival: The Alternative to Extinction* (Garden City: Doubleday & Company, Anchor Books, 1975). See also *The Journal of the New Alchemists* (Woods Hole, Mass.: The New Alchemy Institute, 1973, 1974).

34. One peculiar response of thinkers now worried about the technological society is to pretend in effect that one *already* lives in the future. A sophisticated technology has been directed toward more intelligent ends and given a more humane structure. Through the proper selection of new devices, the problems of the old order have been surmounted. But in these future fantasies, of which "postindustrialism" is now the most popular, there is almost no attempt to

stipulate what will have happened to the technologies we will supposedly have "gone beyond." Since there are new technologies of information processing, for example, we are somehow entitled to assume that the world of industrial technology has vanished.

35. Glenn T. Seaborg and Roger Corliss, *Man and Atom* (New York: E. P. Dutton, 1972), p. 265.

36. See Malcolm I. Thomis, *The Luddites: Machine-Breaking in Regency England* (New York: Schocken Books, 1970); George Rudé, *The Crowd in History* (New York: John Wiley & Sons, 1964), pp. 79-92. I owe the formulation of the Luddite criteria to Larry Spence.

37. Aeschylus, *Prometheus Bound,* in *Aeschylus II,* ed. David Grene and Richard Lattimore and trans. David Grene (New York: Washington Square Press, 1967), pp. 148-149.

38. Ibid., p. 144.

39. Ibid., p. 156.

40. Ibid.

Index